MEI structured mathematics

Statistics 1

ANTHONY ECCLES
ALAN GRAHAM
ROGER PORKESS

MEI Structured Mathematics is supported by industry:
BNFL, Casio, GEC, Intercity, JCB, Lucas, The National Grid Company,
Esso, Texas Instruments, Thorn EMI

Hodder & Stoughton

A MEMBER OF THE HODDER HEADLINE GROUP

British Library Cataloguing in Publication Data

Eccles, Anthony
  Statistics. – Book 1. – (MEI Structured
  Mathematics Series)
  I. Title II. Series
  519.5

  ISBN 0-340-57170-5

First published 1993
Impression number 10 9 8 7
Year   1998   1997

Typeset by Keyset Composition, Colchester, Essex.
Printed in Great Britain for Hodder & Stoughton Educational,
a division of Hodder Headline Plc, 338 Euston Road,
London NW1 3BH by Scotprint Ltd, Musselburgh, Scotland.

MEI Structured Mathematics

Mathematics is not only a beautiful and exciting subject in its own right but also one that underpins many other branches of learning. It is consequently fundamental to the success of a modern economy.

MEI Structured Mathematics is designed to increase substantially the number of people taking the subject post-GCSE, by making it accessible, interesting and relevant to a wide range of students.

It is a credit accumulation scheme based on 45 hour components which may be taken individually or aggregated to give:

  3 Components   AS Mathematics
  6 Components   A Level Mathematics
  9 Components   A Level Mathematics + AS Further Mathematics
12 Components   A Level Mathematics + A Level Further Mathematics

Components may alternatively be combined to give other A or AS certifications (in Statistics, for example) or they may be used to obtain credit towards other types of qualification.

The course is examined by the Oxford and Cambridge Schools Examination Board, with examinations held in January and June each year.

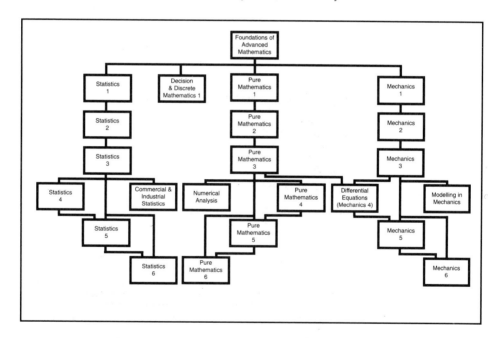

This is one of the series of books written to support the course. Its position within the whole scheme can be seen in the diagram above.

Mathematics in Education and Industry is a curriculum development body which aims to promote the links between Education and Industry in Mathematics at secondary school level, and to produce relevant examination and teaching syllabuses and support material. Since its foundation in the 1960s, MEI has provided syllabuses for GCSE (or O Level), Additional Mathematics and A Level.

For more information about MEI Structured Mathematics or other syllabuses and materials, write to MEI Office, 11 Market Street, Bradford on Avon BA15 1LL.

Introduction

This is the first in a series of books written to support the Statistics Components in MEI Structured Mathematics but you may also use them for an independent course in the subject. Throughout the course the emphasis is on understanding and interpretation, rather than on mere routine calculations.

In this book you meet three fundamental aspects of statistics: exploring data, probability and hypothesis testing.

In the opening chapter, you are shown how to find the most appropriate ways of investigating and displaying sets of data, so that you can discover and communicate their main features. You are encouraged to collect your own figures and get a feel for real data. You then move on to the basic ideas of probability, which underpin all statistical theory. In this course you meet hypothesis testing at an early stage and then learn how to use a steadily increasing number of tests. The first of these, based on the binomial distribution, is in this book and the last three chapters lead you into it.

Several examples are taken from the pages of a fictional local newspaper, *The Avonford Star*. Much of the information that you receive from the media is of a broadly statistical nature. In this book, you are encouraged to recognise this and are shown how to evaluate what you are told.

The authors would like to thank the many people who have helped in the preparation of this book, not least those who read the early versions and trialled them with their students, and also the various examining boards who have given permission for their past questions to be included in the exercises.

Anthony Eccles, Alan Graham and Roger Porkess

Contents

Exploring Data

*A judicious man looks at Statistics,
not to get knowledge but to save himself
from having ignorance foisted on him.*

Carlyle

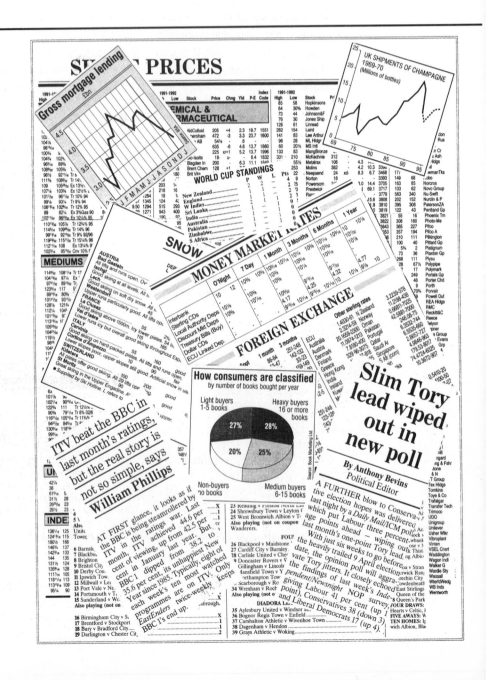

The cuttings on the left all appeared in one newspaper on one day. Some of them give data as figures, others display them or describe them in articles for you the reader.

How do you interpret this information? Which stories do you take seriously and which do you dismiss as being insignificant or even misleading?

To answer these questions fully you need to understand how data are collected and analysed before they are presented to you, and how you should evaluate what you are given to read (or see on the television). That is what the subject of Statistics is all about.

In this book many of the examples are set as stories within a fictitious local newspaper, *The Avonford Star*. Some of them are written as articles; others are presented from the journalists' viewpoint as they sort through data trying to write an interesting story. As you work through the book, look too at the ways you are given such information in your everyday life.

THE AVONFORD STAR

Another cyclist seriously hurt

Will you be next?

On her way back home from school on Wednesday afternoon, little Denise Cropper was knocked off her bicycle and taken to hospital with suspected concussion.

Denise was struck by a Ford Transit van, only 50 yards from her own house.

Denise is the fourth child from the Springfields estate to be involved in a serious cycling accident this year.

The busy road where Denise Cropper was knocked off her bicycle yesterday

After this report the editor of *The Avonford Star* decided that it was time to respond to mounting local worry about cycling accidents involving children. She commissioned one of the paper's reporters to investigate the situation and write a leading article for the paper on it.

The reporter started by collecting data from two sources. He went through back numbers of *The Avonford Star* for the previous two years, finding all the reports of cycling accidents. He also asked an assistant to carry out a survey of the ages of local cyclists; he wanted to know whether most cyclists were children, young adults or whatever.

Before starting to write his article, the reporter needed to make sense of the data for himself. He then had to decide how he was going to present the information to his readers. These are the sorts of data he had to work with.

Name	Age	Distance from home	Cause	Injuries	Treatment
John Smith	45	3 km	skid	Concussion	Hosp. Outpatient
Debbie Lane	5	75 m	hit kerb	Broken arm	Hosp. Outpatient
Arvinder Sethi	12	1200 m	lorry	Multiple	Hosp. 3 weeks
Marion Wren	8	300 m	hit each other	Bruising	Hosp. Outpatient
David Huker	8	50 m		Concussion	Hosp. overnight

There were 32 accidents listed in the reporter's table.

Ages of cyclists (from survey)

66	6	62	19	20		15	21	8	21	63		44	10	44	34	18
35	26	61	13	61		28	21	7	10	52		13	52	20	17	26
64	11	39	22	9		13	9	17	64	32		8	9	31	19	22
37	18	138	16	67		45	10	55	14	66		67	14	62	28	36
9	23	12	9	37		7	36	9	88	46		12	59	61	22	49
18	20	11	25	7		42	29	6	60	60		16	50	16	34	14
18	15														

This information is described as *raw data*, which means that no attempt yet has been made to organise it in order to look for any patterns.

In this chapter you are introduced to a number of techniques for investigating raw data (including these accident statistics). You want to find what information is contained in the data, to decide what are their important features and how they are best displayed.

Looking at the Distribution

At the moment the arrangement of the ages of the 92 cyclists tells you very little at all. The reporter had asked an assistant to collect the information and this was the order in which she presented it.

Clearly these data must be organised so as to reveal the underlying shape, the *distribution*. The figures need to be ranked according to size and preferably grouped as well.

Tally

Tallying is a quick, straightforward way of grouping data into suitable intervals. You have probably met it already.

Stated age A (years)	Tally	Frequency
0 – 9	̶H̶H̶ ̶H̶H̶ ̶l̶l̶l̶	13
10 – 19	̶H̶H̶ ̶H̶H̶ ̶H̶H̶ ̶H̶H̶ ̶H̶H̶ l	26
20 – 29	̶H̶H̶ ̶H̶H̶ ̶H̶H̶ l	16
30 – 39	̶H̶H̶ ̶H̶H̶	10
40 – 49	̶H̶H̶ l	6
50 – 59	̶H̶H̶	5
60 – 69	̶H̶H̶ ̶H̶H̶ llll	14
70 – 79		0
80 – 89	l	1
. . .		
130 – 139	l	1
	TOTAL	92

Outliers

A tally immediately shows up any *outliers*, that is values which are far away from the rest. In this case there are two outliers: 88 and 138. Before doing anything else you must investigate these.

In this case the 88 is genuine, the age of Millie Smith, who is a familiar sight cycling to the shops.

The 138 needless to say is not genuine. It was the written response of a man who was insulted at being asked his age. Since no other information about him is available, this figure is best ignored and the sample size reduced from 92 to 91. You should always try to understand an outlier before deciding to ignore it; it may be giving you important information.

Describing the Shape of a Distribution

An obvious benefit of using a tally is that it shows the overall shape of the distribution.

When the data are drawn as a tally, the variable (in this case age) goes up the vertical axis and the frequency (the number of people) along the horizontal axis. It is more usual to present distributions with the axes the other way round so that these data might then be presented as shown in figure 1.1.

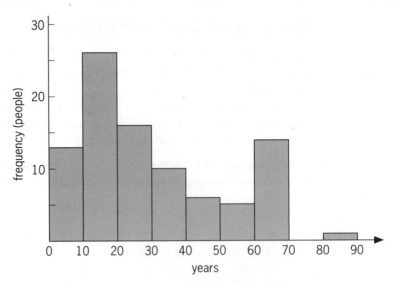

Figure 1.1 Bar chart to show the ages of persons involved in cycling accidents

You can now see at a glance that a large proportion (more than a quarter) of the sample are in the 10 to 19 year age range. This is the *modal* group, as it is the one with the most members. The single value with the most members is called the *mode*, in this case age 9.

You will also see that there is a second peak among those in their sixties; so this distribution is called *bimodal* (or *U-shaped*).

Different types of distribution are described in terms of the position of their modes or modal groups, see figure 1.2.

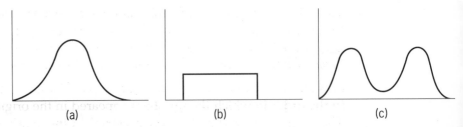

Figure 1.2 Distribution shapes:
(a) symmetrical (mode central); (b) uniform (no mode); (c) U-shaped (bimodal)

When the mode is off to one side the distribution is said to be *skewed*. If the mode is to the left with a long tail to the right the skew is *positive*; if the long tail is to the left the distribution is *negatively skewed*, figure 1.3.

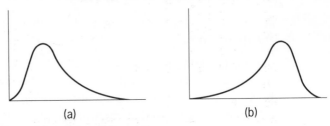

Figure 1.3 Skew: (a) positive; (b) negative

Stem and Leaf Diagrams

The quick and easy view of the distribution from the tally has been achieved at the cost of losing information. You can no longer see the original figures which went into the various groups and so cannot, for example, tell from looking at the tally whether Millie Smith is 80, 81, 82, or any age up to 89. This problem of the loss of information can be solved by using a *stem and leaf diagram*.

This is a quick way of grouping the data so that you can see their distribution and still have access to the original figures. The one below shows the ages of the 91 cyclists surveyed.

stem	
0	6 8 7 9 9 8 9 9 9 7 9 7 6
10	9 5 0 8 3 0 3 7 1 3 7 9 8 6 0 4 4 2 2 8 1 6 6 4 8 5
20	0 1 1 6 8 1 0 6 2 2 8 3 2 0 5 9
30	4 5 9 2 1 7 6 7 6 4
40	4 4 5 6 9 2
50	2 2 5 9 0
60	6 2 3 1 1 4 4 7 6 7 2 1 0 0
70	
80	8

leaves

$n = 91$ 60   7 represents 67 years
Stem and leaf diagram showing the ages of a sample of 91 cyclists (unsorted)

The column of figures on the left (going from **0** to **80**) corresponds to the tens digits of the ages. This is called the *stem* and in this example it consists of 9 levels. Corresponding to each level on the stem are the *leaves* and these represent the units digits of the data values.

As you can see above, the leaves for a particular level have been placed in the order in which the numbers appeared in the original raw data. This is fine for showing the general shape of the distribution, but sometimes it is helpful to sort these numbers into order, as shown below.

0	6 6 7 7 7 8 8 9 9 9 9 9 9
10	0 0 0 1 1 2 2 3 3 3 4 4 4 5 5 6 6 6 7 7 8 8 8 8 9 9
20	0 0 0 1 1 1 2 2 2 3 5 6 6 8 8 9
30	1 2 4 4 5 6 6 7 7 9
40	2 4 4 5 6 9
50	0 2 2 5 9
60	0 0 1 1 1 2 2 3 4 4 6 6 7 7
70	
80	8

This entry corresponds to the only twenty nine year old cyclist in the survey.

$n = 91$ 60   7 represents 67 year
Stem and leaf diagram showing the ages of a sample of 91 cyclists (sorted)

The stem and leaf diagram gives you a lot of information at a glance:

- The youngest cyclist is 6 and the oldest is 88 years of age
- More people are in the 10–19 year age range than in any other 10 year age range
- There are three 61 year olds
- The modal age (i.e. the age with the most people) is 9
- The 17th oldest cyclist in the survey is 55 years of age.

The levels of a stem and leaf diagram do not need to be steps of 10 apart, as in this case. If the same data had been presented in steps of 5, the start of the diagram would have been:

```
 0
 5     1 1 2 2 2 3 3 4 4 4 4 4 4
10     0 0 0 1 1 2 2 3 3 3 4 4 4
15     0 0 1 1 1 2 2 3 3 3 3 4 4
20     0 0 0 1 1 1 2 2 2 3
       etc.
```

Thus the age of each cyclist is the sum of the stem and the entry in the leaf.

Another way of presenting the same figures is as a stem and leaf diagram is like this

```
 0
 0     6 6 7 7 7 8 8 9 9 9 9 9 9
10     0 0 0 1 1 2 2 3 3 3 4 4 4
10     5 5 6 6 6 7 7 8 8 8 8 9 9
20     0 0 0 1 1 1 2 2 2 3
       etc.
```

Exercise 1A

1. Forty motorists entered for a driving competition. The organisers were anxious to know if the contestants had enjoyed the event and also to know their ages, so that they could plan and promote future events effectively. They therefore asked entrants to fill in a form on which they commented on the various tests and gave their ages.

   The information was copied from the forms and the ages listed as:

   ```
   28  52  44  28  38     46  62  59  37  60
   19  55  34  35  66     37  22  26  45   5
   61  38  26  29  63     38  29  36  45  33
   37  41  39  81  35     35  32  36  39  33
   ```

   (i) Plot these data as an unsorted stem and leaf diagram.
   (ii) Identify any outliers and comment on them.

   (iii) Plot a second stem and leaf diagram, this time with the leaves sorted.

   (iv) With which age group was the competition most popular?

   (v) Describe the distribution.

2. The unsorted stem and leaf diagram below gives the ages of males whose marriages were reported in a local newspaper one week.

```
 0
10     9 6 9 8
20     5 6 8 9 1 1 0 3 6 8 4 1 2 7
30     0 0 5 2 3 9 1 2 0
40     8 4 7 9 6 5 3 3 5 6
50     2 2 1 7
60
70
80     3
```

n = 42   **30**   2 represents 32

   (i) Identify and comment on any outliers.

   (ii) What was the age of the oldest person whose marriage is included?

   (iii) Redraw the stem and leaf diagram with the leaves sorted.

   (iv) Redraw the stem and leaf diagram with steps of 5 years between the levels rather than 10.

   (v) Describe and comment on the distribution.

3. The following marks (which are genuine) were obtained on an A Level Mathematics paper by the candidates at one centre.

```
26 54 50  37 54    34 34 66 44 76    45 71  51 75 30
29 52 43  66 59    22 74 51 49 39    32 37  57 37 18
54 17 26  40 69    80 90 95 96 95    70 68  97 87 68
77 76 30 100 98    44 60 46 97 75    52 82  92 51 44
73 87 49  90 53    45 40 61 66 94    62 39 100 91 66
35 56 36  74 25    70 69 67 48 65    55 64
```

Draw a stem and leaf diagram to illustrate these marks and comment on their distribution.

4.   **Ages of a sample of 40 hang-gliders (years)**

```
28 19 24 20 28    26 22 19 37 40    19 25 65 34 66
35 69 65 26 17    22 26 45 58 30    31 58 26 29 23
72 23 21 30 28    65 21 67 23 57
```

   (i) Using intervals of 5 years, draw a stem and leaf diagram to illustrate these figures.

   (ii) Comment on and explain the shape of the distribution.

5. An experimental fertiliser called GRO was applied to 50 lime trees, chosen at random, in a plantation. Another 50 trees were left untreated. The yields in kg were as follows:

Treated

59 25 52 19 32	26 33 24 35 30	23 54 33 31 25
23 61 35 38 44	27 24 30 62 23	47 42 41 53 31
20 21 41 33 35	38 61 63 44 18	53 38 33 49 54
50 44 25 42 18		

Untreated

8 11 22 22 20	5 31 40 14 45	10 16 14 20 51
55 30 30 25 29	12 48 17 12 52	58 61 14 32 5
29 40 61 53 22	33 41 62 51 56	10 48 50 14 8
63 43 61 12 42		

Draw stem and leaf diagrams to compare the two sets of data and comment on the effects of GRO.

6. A group of 25 people were asked to estimate the length of a line which they were told was between 1 and 2 metres long. Here are their estimates, in metres.

1.15 1.33 1.42 1.26 1.29	1.30 1.30 1.46 1.18 1.24
1.21 1.30 1.32 1.33 1.29	1.30 1.40 1.26 1.32 1.30
1.41 1.28 1.65 1.54 1.14	

(i) Represent these data in a sorted stem and leaf diagram.
(ii) Would you consider any of the figures to be outliers?
(iii) From the stem and leaf diagram which you drew, read off the third highest and third lowest length estimates.
(iv) Find the middle of the 25 estimates.
(v) On the evidence that you have, could you make an estimate of the length of the line? Justify your answer.

7. The IQs of the members of a group of trainees on a course were measured as follows:

112 119 136 125	127 144 131 124	137 114 133 122
132 132 117 122	143 114 129 124	140 113 138 141
126 117 118 136		

Their tutor asked not to be shown the results in case they influenced the way he treated the trainees. At the end of the course he gave them an examination; the marks of the trainees (not given in the same order as the figures above) were:

62 24 27 82 96	33 67 60 90 25	79 93 61 83 31
23 59 100 72 65	75 76 41 38 70	31 76 57

(i) Draw stem and leaf diagrams for the two distributions.

Statistics 1

(ii) Compare the two diagrams and comment on what you find. Do you think the examination was too hard? Do you think the tutor was pitching his lessons at the correct level for the group? What assumptions have you made about the measurement of IQ?

8. The table below gives the annual rates of inflation for 12 countries in 1992 and 1991.   13

Country	Inflation, annual change, %	
	1992	**1991**
UK	4.1	8.9
Australia	1.5	6.9
Belgium	2.3	3.9
Canada	1.6	6.8
France	2.9	3.4
Germany	4.0	2.8
Italy	6.1	6.5
Japan	2.2	3.3
Netherlands	4.1	2.8
Spain	5.5	6.6
Sweden	7.9	10.9
USA	2.6	5.7
OECD	5.6	6.7

(Source: *The Independent on Sunday* 1/3/92)

(i) Copy and complete this back-to-back stem and leaf diagram and comment on your results.

1992			**1991**
5	1.0		
	2.0		
	3.0		
1	4.0		
	5.0		
	6.0	9	
	7.0		
	8.0	9	
	9.0		
	10.0		

(ii) What loss of information does this diagram involve?

(iii) Can you think of an alternative representation which deals with this problem? Do you think a stem and leaf diagram is a good way of presenting these data?

9. Construct a stem and leaf diagram with the following features:

    (i)   the range from lowest to highest values is 56
    (ii) the modal group is $30 - 39$
    (iii) the sample size, n, is 25
    (iv) the median (i.e. the middle value) is 27
    (v)   the highest value is 65
    (vi) the width of the intervals is 10.

10. Construct a stem and leaf diagram with the following features:

    (i)   the range from lowest to highest values is 28
    (ii) the two modal groups (which have an equal frequency of 8) are $60 - 64$ and $65 - 69$
    (iii) the sample size, n, is 30
    (iv) the median (i.e. the middle value) is 67
    (v)   the highest value is 85
    (vi) the width of the intervals is 5.

Grouping Data

Grouping means putting the data into a number of *classes*. The number of items of data falling into any class is called the *frequency* for that class.

Categorical Data

Some data comes to you already in classes or categories. Such data, like these for the results of the 1992 General Election for the constituencies in Wales, are called *categorical*.

L L L C L   PC L L C L   L L PC C L   C L L L L
L PC L C LD   L L L L L   L L L L L   C L PC

C – Conservative; L – Labour; LD – Liberal Democrat; PC – Plaid Cymru (results listed alphabetically, by constituency).

When collected together this gives

Wales	Conservative	Labour	Liberal Democrat	Plaid Cymru
MPs	6	27	1	4

There are several ways of displaying this sort of data including a *pie chart*, a *vertical line chart* and a *bar chart*.

When pie charts are being used to contrast similar information from two or more sources, the area rather than the radius of each should be proportional to the total frequency it represents.

The equivalent figures for Scottish MPs after the 1992 General Election are:

Scotland	Conservative	Labour	Liberal Democrat	Nationalist (SNP)
MPs	11	49	9	3

In the pie charts (figure 1.4a), that for the 38 Welsh MPs has radius 2 cm and so area 4π. There are 72 Scottish MPs and so their pie chart has area

$$\frac{72}{38} \times 4\pi.$$

This means that its radius, R, is given by

$$\pi R^2 = \frac{72}{38} \times 4\pi,$$

and so has value 2.75 cm (figure 1.4b).

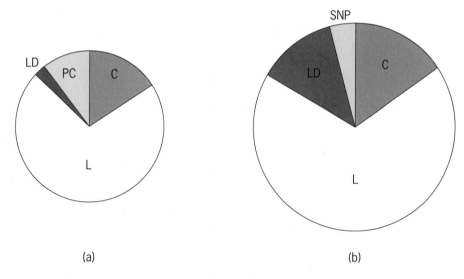

(a)

(b)

Figure 1.4

There are many different ways of drawing bar charts, with the bars horizontal or vertical, as vertical lines (a vertical line chart, figure 1.5), or subdivided (a compound bar chart, figure 1.6). Often there is no single right way of displaying the information; what is most important is that it should be easy to follow and not misleading.

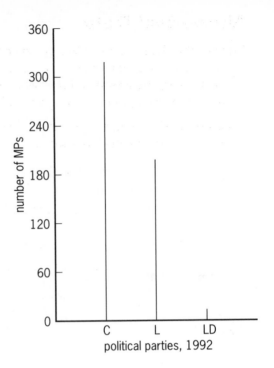

Figure 1.5 Vertical line chart showing MPs for English constituencies (1992)

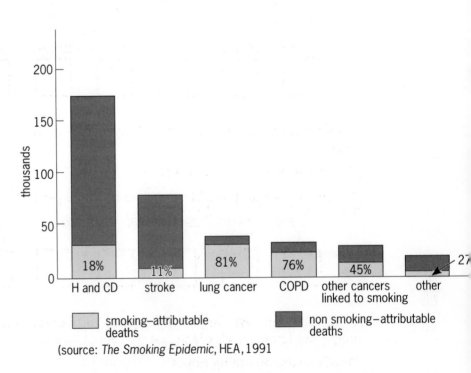

(source: *The Smoking Epidemic*, HEA, 1991

Figure 1.6 Compound bar chart showing smoking- and non-smoking-attributable deaths

Numerical Data

Most of the data you meet will however be grouped according to the value of a *variable*, like people's heights or weights, or the score on a die. In such cases, the classification is done on the basis of a number, rather than a description, and the data are called *numerical*.

When numerical data are grouped, each item of data falls within a *class interval* lying between *class boundaries*.

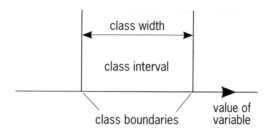

Figure 1.7

The length of a class interval is called the class width, see figure 1.7.

You must always be careful about the choice of class boundaries because it must always be absolutely clear to which class any item belongs. A form with the following wording:

Tick one of these boxes for your age:

0 – 10	10 – 20	20 – 30	30 – 40	40 – 50	50+
☐	☐	☐	☐	☐	☐

would cause problems. A 10 year old could tick either of the first two boxes.

A better form of wording would be:

Tick one of these boxes for your age in completed years:

0 – 9	10 – 19	20 – 29	30 – 39	40 – 49	50 or more
☐	☐	☐	☐	☐	☐

Notice that this says "in completed years". Otherwise a 9½ year old might not know which of the first two boxes to tick.

Another way of writing this is:

$0 \leq A < 10$ $10 \leq A < 20$ $20 \leq A < 30$ $30 \leq A < 40$ $40 \leq A < 50$ $50 \leq A$

Even someone aged 9 years and 364 days would clearly still come in the first group.

Discrete and Continuous Data

Before grouping data you must be clear whether the variable they refer to is discrete or continuous, and if the latter, how they have been rounded.

The number of goals a football team scores in a match is 0, 1, 2 or . . . Such data which take only particular values are said to be *discrete*. Other examples of discrete data are the score on a die, the number of letters in a word, the number of coaches in a train etc. Note that discrete data can take fractional values (0, ½, 1, . . .) or negative values (−1, −2, −3, . . .). The important point is that for discrete data all possible values can be listed.

Variables like distance, weight and temperature are *continuous* because if measured accurately enough they can take any values and do not change in particular steps. It would not be possible to list all the values which someone's height might take.

One of the examples we have used so far, that of age, is rather a special case since it is nearly always given rounded down (ie truncated). Although your age changes continuously every moment of your life, you actually state it in steps of one year, in completed years, and not to the nearest whole year. So a man who is a few days short of his 40th birthday will happily say he is 39.

In practice, data for a continuous variable are always given in a rounded form

- a person's height, h, is given as 168 cm, measured to the nearest cm; $167.5 \leq h < 168.5$
- a temperature, T, is given as 21.8°C, measured to the nearest tenth of a degree; $21.75 \leq T < 21.85$
- the depth of an ocean, d, is given as 9200 m, measured to the nearest 100 m; $9150 \leq d < 9250$.

(Notice the rounding convention used here: that if a figure is on a borderline it is rounded up. There are alternative rounding conventions.)

There are usually no problems in grouping discrete data. It is quite clear what any particular value means. On the other hand you have to be very careful when grouping data for a continuous variable. The data below refer to the diameters of trees in a hardwood forest in cm:

$$105 \quad 65 \quad 90 \quad 45 \quad 55 \qquad 135 \quad 85 \quad 75 \quad 80 \quad 95$$
$$65 \quad 70 \quad 75 \quad 65 \quad 90 \qquad 70 \quad 70 \quad 75 \quad 105 \quad 60$$

You can see at once by looking at the figures that they have all been rounded to the nearest 5 cm. A tree of diameter 92 cm would be recorded as 90 cm whereas one of 92.8 cm would be recorded as 95 cm.

Consequently data grouped together under the heading $90 \leq d < 100$ would refer to trees recorded as 90 and 95 but the diameters of the actual trees would be between 87.5 and 92.5 (recorded as 90) and 92.5 and 97.5 (recorded as 95). So they would really lie between 87.5 and 97.5 cm. Data that are, as accurately as can be measured, on a boundary are usually rounded up, so this interval is written as $87.5 \leq d < 97.5$.

So be careful!

Histograms

The data the reporter collected when researching his article on cycling accidents included the distances from home of those involved in cycling accidents. In full these were as follows, in metres:

3000	75	1200	300	50	10	150	1500	250	25
200	4500	35	60	120	400	2400	140	45	5
1250	3500	30	75	250	1200	250	50	250	450
15	4000								

You can see at a glance that there is a considerable spread within these figures to the extent that it would be almost impossible to draw a sensible stem and leaf diagram with them.

It would be much better first to group them conveniently and then to display them as a *histogram*. One possible grouping is:

Location relative to home	Distance, d,	Frequency (accidents)
Very close	$0 \leq d < 100$	12
Close	$100 \leq d < 500$	11
Not far	$500 \leq d < 1500$	3
Quite far	$1500 \leq d < 5000$	6

Notice that although there has clearly been some rounding in collecting the original data, it is not really possible to allow for it in setting the class boundaries in this case, because we don't know precisely how the rounding was done.

The next step is to work out the *frequency density* for each class. You do this by dividing each frequency by the class width, as follows:

Class Interval (metres)	Frequency (accidents)	÷	Class width (metres)	=	Frequency density (accidents/metre)
$0 \leq d < 100$	12	÷	100	=	0.12
$100 \leq d < 500$	11	÷	400	=	0.0275
$500 \leq d < 1500$	3	÷	1000	=	0.003
$1500 \leq d < 5000$	6	÷	3500	=	0.0017

You will notice that the numbers to be used on the vertical scale (the frequency density) are all much less than 1. That is because we calculated the frequency density in accidents/metre and this is clearly not the most convenient unit to use.

If instead we took 100 m as being a unit for the class widths, the table would then become:

Class Interval (metres)	Frequency (accidents)	÷	Class width (100 metres)	=	Frequency density (accidents/100 metres)
$0 \leq d < 100$	12	÷	1	=	12.00
$100 \leq d < 500$	11	÷	4	=	2.75
$500 \leq d < 1500$	3	÷	10	=	0.30
$1500 \leq d < 5000$	6	÷	35	=	0.17

The resulting histogram is then drawn, figure 1.8.

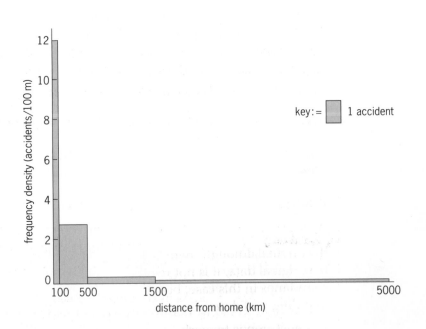

Figure 1.8

You will see at once that the rectangle representing $0 \leq d < 100$ is the highest and that tells you that this is the *modal class* – that with the highest frequency density.

In simple English that means that most accidents occur near home. Can you say why this is likely to be the case?

The main points to remember when drawing a histogram are:

- Histograms are usually used for representing continuous data. For discrete data it is more common to draw a barchart (but you may sometimes see such a diagram for discrete data called a histogram).

- Because the data are continuous, adjacent columns of the histogram should touch (unlike a barchart where the columns are often drawn with gaps between the columns).

- It is the *areas* and not the heights of the columns that give the frequency of each interval. The last column of this one, for example, has area $35 \times 0.17 = 6$, the frequency of accidents for that interval. (Note that 0.17 is a rounded figure.)

- The vertical axis should be marked with frequency density, rather than a frequency. There should be a box beside the histogram as a *legend*, showing the unit area.

CAUTION

The term 'histogram' is sometimes also used to describe a diagram displaying data from a continuous distribution with equal class intervals but with frequency rather than frequency density on the vertical scale. While this is not technically correct, it is done sufficiently often for the term histogram to have become accepted as a description for that sort of diagram also.

Exercise 1B

1. State (i) whether the data described below are categorical or numerical, and, if numerical, whether discrete or continuous

(ii) what you think would be the most appropriate method of displaying the data

the number of coins in shoppers' purses

the colour of the eyes of a sample of people

the weights of a sample of eggs

the medals (gold, silver and bronze) won by a team at the Olympic games

the sizes of a sample of eggs (size 1, size 2, etc)

the times of the runners in a 100 m race

the numbers on the shirts of a sample of rugby players

the scores of the competitors in an iceskating competition

estimates of the length of a needle

the single (year) letter on the registration plate of a sample of cars.

Exercise 1B continued

2. The two pie charts are drawn to scale. They represent the income of two North Atlantic islands one year. The total income for Seanna was £72 000. Calculate the earnings for both islands from each product.

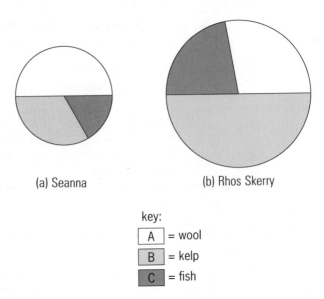

(a) Seanna   (b) Rhos Skerry

key:
| A | = wool
| B | = kelp
| C | = fish

3. The compound bar chart shows the production of three cash crops from a region of an African country in the years 1980 and 1990.

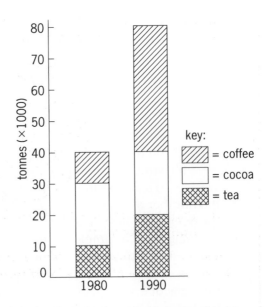

key:
= coffee
= cocoa
= tea

(i)   Draw, to scale, two pie charts to illustrate the same information.
(ii) Comment on the changes in crop production over the 10 years.
(iii) Which do you consider to be the better method of display and why?

4. A developing country's education budget is divided between the Primary, Secondary and Tertiary sectors. The figures below give the amount of money spent on each sector. They are in millions of dollars and have been adjusted to allow for inflation.

	Primary	Secondary	Tertiary
1960	4.1	1.1	0.9
1975	12.2	8.4	6.0
1990	18.0	20.8	16.9

(i)   Draw compound bar charts to illustrate this information.
(ii) Draw, to scale, three pie charts to illustrate the information.
(iii) Comment briefly on what the figures tell you about the country's education programme.
(iv) Which do you consider to be the better method of display and why?

5. The status of the 120 full-time employees of a small factory is linked to their pay.

Status	Pay (£*P*/week)
Unskilled	$120 \leq P < 200$
Skilled	$200 \leq P < 280$
Staff	$280 \leq P < 400$
Management	$400 \leq P < 1000$

The company's personnel department illustrate the numbers in the various groups on this accurately drawn pie chart.

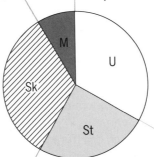

(i)   Copy and complete this table, finding the frequencies from the pie chart

Status	Interval (pay £*P*/week)	Frequency (employees)	÷	Interval Length (× £10)	=	Frequency Density (employees/£10)
Unskilled	$120 \leq P < 200$	40	÷	8	=	5.0
Skilled	$200 \leq P < 280$					
Staff	$280 \leq P < 400$					
Management	$400 \leq P < 1000$					

Exercise 1B continued

(ii) Hence draw a histogram to illustrate the data.

(iii) Is the shape of the histogram about what you would expect?

(iv) Which do you think is the more appropriate method of display?

6. A number of trees in two woods were measured. Their diameters, correct to the nearest centimetre, are summarised in the table below.

Diameter (cm)	1–10	11–15	16–20	21–30	31–50	Total
Akeley Wood	10	5	4	10	1	30
Shaws Wood	6	8	20	5	1	40

(Trees less than ½ cm in diameter are not included.)

(i)   Write down the actual class boundaries.

(ii) Draw two separate histograms to illustrate this information.

(iii) State the modal class for each wood.

(iv) Describe the main features of the distributions for the two woods.

7.   *UK Motorcycle casualties, by age, 1989*

Age (years)	% age casualties
16–19	36
20–24	26
25–29	15
30–39	11
40–80	12

(Source: *Social Trends* 21)

(i)   Draw a pie chart to illustrate these data, using the five groups given.

(ii) Write down the ages of the youngest and oldest person who would come within the interval written as 16–19.

(iii) Draw a histogram to represent these data.

(iv) Comment on the information in the table, stating with reasons what you consider to be the most and the least dangerous ages for motorcyclists. Do you have enough information to answer this as fully as you would wish?

(v)   Which did you find the more helpful method of display, the pie chart or the histogram?

8   Listed below are the prime numbers, p, from 1 up to 1000. (1 itself is not usually defined as a prime.)

Primes up to 1000

2	3	5	7	11	13	17	19	23	29	31	37	41	43
47	53	59	61	67	71	73	79	83	89	97	101	103	107
109	113	127	131	137	139	149	151	157	163	167	173	179	181
191	193	197	199	211	223	227	229	233	239	241	251	257	263
269	271	277	281	283	293	307	311	313	317	331	337	347	349
353	359	367	373	379	383	389	397	401	409	419	421	431	433
439	443	449	457	461	463	467	479	487	491	499	503	509	521
523	541	547	557	563	569	571	577	587	593	599	601	607	613
617	619	631	641	643	647	653	659	661	673	677	683	691	701
709	719	727	733	739	743	751	757	761	769	773	787	797	809
811	821	823	827	829	839	853	857	859	863	877	881	883	887
907	911	919	929	937	941	947	953	967	971	977	983	991	997

(i) Draw a histogram to illustrate these data with the following class intervals:

$1 \leqslant p < 20$ $20 \leqslant p < 50$ $50 \leqslant p < 100$ $100 \leqslant p < 200$ $200 \leqslant p < 300$ $300 \leqslant p < 500$ and $500 \leqslant p < 1000$.

(ii) Comment on the shape of the distribution.

9. Buy two newspapers on the same day, one tabloid and the other from the quality press. Find a story which is covered in both of them. Record the lengths of the first 100 words in each of the articles, and display them appropriately.

Are there differences between the two papers which can be seen at a glance from your display?

Now do a survey of the lengths of the sentences in the two papers. Are there differences which can be seen at once when the data are suitably displayed?

10. After completing a long assignment, a student was told by his tutor that it was more like a book than an essay. He decided to investigate how many pages there are in a typical book and started by writing down the numbers of pages in the books on one of his shelves, as follows:

256	128	160	128	192	64	356	96	64	160
464	128	96	96	556	148	64	192	96	512
940	676	128	196	640	44	64	144	256	72

(i) Look carefully at these data and state, giving your reasons, whether they are continuous or discrete. Give an explanation for your answer.

(ii) Decide on the most helpful method of displaying these data and draw the appropriate diagram.

Measures of Central Tendency: Mean, Mode, Median, Midrange

What was the typical age of the people who responded to the questionnaire on cyclists' ages? What was the typical distance from home of someone involved in an accident?

To answer such questions you need some sort of *average*, or typical value. There are several measures you can use for this.

Mean

The *mean*, $\bar{x}$, is the sum of all the data values divided by the number of them. This is often referred to in everyday English as the average.

For the cyclists' ages, Mean $= 2717 \div 91 = 29.9$ years.
For the distances of accidents from home, Mean $= 25\,785 \div 32 = 806$ m.

The mean is the most commonly used form of average in statistics.

NOTES

1. *The mean of the cyclists' ages needs to be adjusted because age is always rounded down. The mean error produced by this is 0.5 year so this should be added to the figure of 29.9 years to give 30.4 years.*
2. *The mean described here is correctly called the* arithmetic mean; *there are other forms (geometric mean, harmonic mean, weighted mean) all of which have particular applications.*

Mode

The *mode* is the value which occurs most frequently. If two values occur more frequently than the rest, the distribution is said to be *bimodal*, even if the frequencies are not the same for both modes.

When data are grouped the *modal group* is that with the largest frequency density, and so has the highest rectangle in a histogram; if all the groups are of equal class width the modal group is that with the greatest frequency.

For the cyclists' ages (see histogram on page 6), the mode is 9 years (the frequency is 6). The modal group (for classes of width 10 years) is $10-19$ years (the frequency is 26). There is however a second mode in the $60-69$ range so the distribution is bimodal.

For the distances of accidents from home, the mode is 250 m (the frequency is 4). The modal group (for the classes given earlier) is $0-100$ m (the frequency density is 12 accidents/100 m).

For a small set of discrete data the mode can often be misleading, especially if there are many values the data can take, with a lot of data happening to fall on a particular value. With grouped data, however, the modal group or groups will nearly always provide valuable information.

Median

The *median* is the value of the middle item when all the data items are ranked in order. If there is an even number of items, there is no single middle one and so a value midway between those either side of the middle is taken.

For the ages of 91 cyclists, the median is the age of the 46th person which is 22 years.

For the data for the 32 accidents (see page 53) there is no single middle item. The 16th has value 250 m, the 17th 200 m, so the median is placed half way between these at 225 m.

For n items of data, the median is that located at position $\frac{1}{2}(n + 1)$ if n is odd; if n is even, it is the mean of the values at positions $\frac{1}{2}n$ and $\frac{1}{2}n + 1$.

The median is easy to work out if the data are already ranked, otherwise it can be very tedious. It usually provides a good representative value and is much less influenced by outliers (if at all) than mean or midrange.

Midrange

The *midrange* is the value midway between the upper extreme and the lower extreme values in the data set.

For the cyclists' ages the midrange is $\frac{1}{2}(6 + 88) = 47$ years, for the distances of the accidents it is $\frac{1}{2}(5 + 4500) = 2252.5$ m.

The midrange is easy to calculate but is only useful if the data are reasonably symmetrical and free from outliers. A single outlier can have a major effect on the midrange; the midrange of the cyclists' ages is increased by $10\frac{1}{2}$ years by the inclusion of the 88 year old, Millie Smith.

For Discussion

Which of these averages is the most suitable to use depends on the particular data you are using at any time and why you are using it. Which do you think are the most useful in these two examples? The information is summarised for you below.

	Cyclists' ages	Distances from home
Mean	$2717 \div 91 = 29.9$ years	$25785 \div 32 = 806$ m
Mode	9 years	250 m
Modal group	10 – 19 years	0 – 100 m
Median	22 years	$\frac{1}{2}(200 + 250) = 225$ m
Midrange	$\frac{1}{2}(6 + 88) = 47$	$\frac{1}{2}(5 + 4500) = 2252.5$ m

The Reporter's Article

The reporter decided that he had enough information and wrote the article below. It is a fair article? Is it justified, based on the available evidence?

THE AVONFORD STAR

A Council that does not care

The level of civilisation of any society can be measured by how much it cares for its most vulnerable members.

On that basis our Town Council rate somewhere between savages and barbarians. Every day they sit back complacently while those least able to defend themselves, the very old and the very young, run the gauntlet of our treacherous streets.

I refer of course to the lack of adequate safety measures for our cyclists, 60% of whom are children or senior citizens. Statistics show that they only have to put one wheel outside their front doors to be in mortal danger. 90% of cyc-

ling accidents happen within 1 mile of home.

Last week Denise Cropper became the latest unwitting addition to these statistics. Luckily she is now on the road to recovery but that is no thanks to our unfeeling council who set people on the road to death and injury without a second thought.

What, this paper asks our councillors, are you doing about providing safe cycle tracks from our housing estates to our schools and shopping centres? And what are you doing to promote safety awareness among our cyclists, young and old?

Answer: Nothing.

Finding the Mean of Grouped Data

When data have been grouped, the original values may have been lost. In that case you would have to calculate the mean by assuming that all the values in an interval are equally spaced about its midpoint. Such a calculation will of course only give you an estimate, not an accurate answer.

EXAMPLE

As part of a research project, Jasmine catches and weighs 80 cod. She records their weights, as follows:

Weight (w kg)	$w < 1$	$1 \leq w < 2$	$2 \leq w < 3$	$3 \leq w < 4$	$4 \leq w < 5$
Frequency	2	23	31	19	5

Calculate an estimate of the mean weight of the fish.

Solution

The calculation is shown in the table below:

Interval	Midpoint	× Frequency	=	Total Weight
$0 < w < 1$	0.5	× 2	=	1.0
$1 \le w < 2$	1.5	× 23	=	34.5
$2 \le w < 3$	2.5	× 31	=	77.5
$3 \le w < 4$	3.5	× 19	=	66.5
$4 \le w < 5$	4.5	× 5	=	22.5
		Total 80		202.0

$$\text{Mean} = 202.0 \div 80 = 2.525 \text{ kg}$$

Exercise 1C

1. Find the mode, mean, midrange and median of these figures:

   (a) 23 46 45 45 29   51 36 41 37 47   45 44 41 31 33

   (b) 110 111 116 119 129   126 132 116 122 130
       116 132 118 122 127   132 126 138 117 111

   (c) 5 7 7 9 1   2 3 5 6 6   8 6 5 7 9   2 2 5 6 6
       6 4 7 7 6   1 3 3 5 7   8 2 8 7 6   5 4 3 6 7

2. For each of these sets of data
   (i) Find the mode, mean, midrange and median.
   (ii) State with reasons which you consider to be the most
        appropriate form of average to describe the distribution.

   (a) The ages of children in a class in years and months:

       14.1 14.11 14.5 14.6 14.0   14.7 14.7 14.9 14.1   14.2
       14.6 14.5   14.8 14.2 14.0   14.9 14.2 14.8 14.11 14.8
       15.0 14.7   14.8 14.9 14.3   14.5 14.4 14.3 14.6   14.1

   (b) The shoe sizes of children in a class:

        3   2½ 5 4 3½   4   4 4 2½ 6   4½ 5 5½ 4 3½
        1½ 3 3 4 2½   3½ 5 4 3   4½   3½ 4 5   3½ 4½

   (c) The number of pints of beer drunk by people in *The Crown and
       Anchor* one Friday evening:

       4 0 0 0 ½   5 3 4 0 0   1½ 0 4 8 0
       4 4 ½ 0 6   3 3 4 5 4   ½ 3 0 4 4

   (d) Students' marks on an examination paper:

       55 78 45 54 0   62 43 56 71 65   0 67 75 51 100
       39 45 66 71 52   71 0 0 59 61   56 59 64 57 63

   (e) The scores of a cricketer during a season's matches:

       10 23 65 0 1   24 47 2 21 53   5 4 23 169 21
       17 34 33 21 0   10 78 1 56 3   2 0 128 12 19

Exercise 1C continued

(f) Scores when a die is thrown 40 times:

2 4 5 5 1   3 4 6 2 5   2 4 6 1 2   5 4 4 1 1
3 4 6 5 5   2 3 3 1 6   5 4 2 1 3   3 2 1 6 6

3. (i) Find the mode, mean, midrange and median of:

2 8 6 5 4   5 6 3 6 4   9 1 5 6 5

Hence write down, without further working, the mode, mean, midrange and median of:

(ii) 20 80 60 50 40   50 60 30 60 40   90 10 50 60 50

(iii) 12 18 16 15 14   15 16 13 16 14   19 11 15 16 15

(iv) 4 16 12 10 8   10 12 6 12 8   18 2 10 12 10

4. The tally below gives the scores of the football teams in the matches of the 1982 World Cup finals.

```
0    ЖТ  ЖТ  ЖТ  ЖТ  ЖТ  ЖТ  I
1    ЖТ  ЖТ  ЖТ  ЖТ  ЖТ  ЖТ  ЖТ  III
2    ЖТ  ЖТ  ЖТ  I
3    ЖТ  III
4    ЖТ  I
5    II
6
7
8
9
10   I
```

(i) Find the mode, mean, midrange and median of this data.
(ii) State which of these you think is the most representative measure.

5. The vertical line chart shows the number of times the various members of a school year had to take their driving test before passing it.

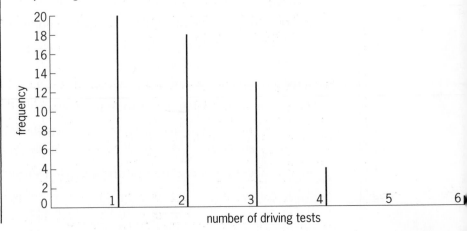

number of driving tests

    (i) Find the mode, mean, midrange and median of this data.

    (ii) State which of these you think is the most representative measure.

6. The figures in the table below are the ages, to the nearest year, of a random sample of 30 people negotiating a mortgage with a bank.

29 26 31 42 38   36 39 49 40 32   33 31 33 52 44
45 35 37 38 38   32 34 27 61 29   32 30 38 42 33

Copy and complete the following stem and leaf diagram. Use the diagram to identify *two* features of the shape of the distribution.

25	4 1
30	1
35	

Find the mean age of the 30 people. Given that 18 of them are men and that the mean age of the men is 37.72, find the mean age of the 12 women.

[MEI]

7. The lengths of time of sixty songs recorded by a certain group of singers are summarised in the table below:

Song length in seconds (x)	Number of songs
$0 < x \leq 120$	1
$120 < x \leq 180$	9
$180 < x \leq 240$	15
$240 < x \leq 300$	17
$300 < x \leq 360$	13
$360 < x \leq 600$	5

    (i) Display the data on a histogram.

    (ii) Determine the mean song length, giving your answer to the nearest half second.

[C]

8. A teacher drives to school every day through a town where the traffic conditions are variable. Each day she makes a note of her journey time to the nearest minute, and at the end of the term she tabulates the information with the following result:

Time in minutes	8	9	10	11	12	13	14	15
Number of journeys	2	3	10	15	18	8	3	1

    (a) Calculate the mean time for the journey.

Exercise 1C continued

(b) There is an alternative route, longer but with less traffic; if the teacher travels by this route, she can be sure that the journey will always take 12½ minutes. Calculate the total extra time she would have spent on travelling if she had used the alternative route every day.

[L]

9. A case containing 270 oranges was opened and each orange was weighed. The masses were found to be distributed as in the following table:

Mass (grammes)	Number of oranges
60 – 100	20
100 – 120	60
120 – 140	80
140 – 160	50
160 – 220	60

Draw a histogram to illustrate the data.
From the table, calculate an estimate of the mean mass of an orange from this crate.

[MEI]

10. In an agricultural experiment, 320 plants were grown on a plot, and the lengths of the stems were measured ten weeks after planting. The lengths were found to be distributed as in the following table:

Length (cm)	Number of plants
20 – 32	30
32 – 38	80
38 – 44	90
44 – 50	60
50 – 68	60

On squared paper, draw a histogram to illustrate the data.
From the table, calculate an estimate of the mean length of stem of a plant from this experiment.

[MEI]

Measures of spread

In the last section you saw how different types of average are used to find the typical value of a data set. Often you are also interested in the *spread* of the data and in this section we will look at some ways to measure this.

The figures below are the prices, in pence, of a 100 gm jar of Nesko coffee in ten different shops:

161 161 163 163 167   168 170 172 172 172

Here are a number of measures of spread for this data.

Range

The simplest measure of spread is the *range*, which is just the difference between the largest value in the data set (the upper extreme) and the smallest value (the lower extreme).

$$\text{Range} = \text{Largest} - \text{Smallest}$$

In this case $\text{Range} = 172 - 161 = 11\,\text{p}.$

NOTE

Be careful not to confuse the range with the midrange. The midrange for these data is 166.5 p.

Interquartile Range

The median is the value of the middle item when all the data items have been ranked in order, or the mean of the two middle values for a data set with an even number of items. The values one quarter of the way through the data set and three quarters of the way through the data set are called the *lower quartile* and the *upper quartile*. They are also called the first and third quartiles, Q_1 and Q_3 respectively, and the median is the second quartile, Q_2.

Quartiles are used mainly with large data sets and their values found by looking at the ¼, ½ and ¾ points. So for a set of size 1000, you would take Q_1 to be at the 250th data point, Q_2 at 500th and Q_3 at 750th. If possible you should avoid using quartiles with small data sets, or samples, since the value of any of them is heavily dependent on the values of one or two members of the set. If you cannot avoid working with a small sample use the values at the positions $\frac{1}{4}n + \frac{1}{2}$, $\frac{1}{2}n + \frac{1}{2}$, $\frac{3}{4}n + \frac{1}{2}$. Since these are often not whole numbers, the calculation may involve interpolation between the values in the sample. In the case of the coffee jars, this gives:

161	161	163	163	167	168	170	172	172	172
1	2	3	4	5	6	7	8	9	10

Q_1 has position
$\frac{1}{4} \times 10 + \frac{1}{2} = 3$
value 163

Q_2 has position
$\frac{1}{2} \times 10 + \frac{1}{2} = 5\frac{1}{2}$
value 167½

Q_3 has position
$\frac{3}{4} \times 10 + \frac{1}{2} = 8$
value 172

NOTE

Many sources state the position of quartiles to be $\frac{1}{4}(n+1)$, $\frac{1}{2}(n+1)$, $\frac{3}{4}(n+1)$. However these points do not partition the data set equally and lead to values for Q_1 and Q_3 which are in most cases more extreme than the true values in the underlying population. The points given above do partition the data set equally and usually lead to values of Q_1 and Q_3 which are substantially closer to the true values.

The difference between the upper and the lower quartile is called the *interquartile range*.

$$\text{Interquartile range} = Q_3 - Q_1$$

In the case of the coffee prices this is $172 - 163 = 9$ p.

The interquartile range covers the middle half of the data set and is a measure of the difference between a typical high value and a typical low value.

The *semi-interquartile range* is also used sometimes. It is half the value of the interquartile range. In the example above, the semi-interquartile range is ½ of 9 pence which is 4.5 pence.

Semi-interquartile range is a measure of the distance of a typically high or low value above or below the middle.

Box-and-Whisker plots (boxplots)

The three quartiles and the two extremes of a set of data may be illustrated in a *box-and-whisker plot*. This is designed to give an easy-to-read representation of the location and spread of a distribution.

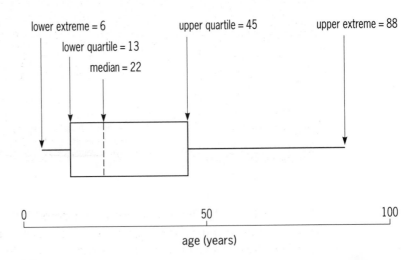

Figure 1.9

The boxplot, figure 1.9, shows these main features for the ages of the 91 cyclists in the survey at the start of the chapter. You will see that it consists of a *box*, representing the middle half of the data (between the quartiles), and two *whiskers* showing the more extreme values, above and below the quartiles. Look at it carefully and decide what you can deduce about the overall shape of this particular distribution.

Percentiles

Just as the quartiles divide up the data set into quarters, *percentiles* divide it into hundredths. Thus, the median corresponds to the 50th percentile, the lower quartile to the 25th percentile and the upper quartile the 75th percentile. In the same way *sextiles* divide up the data into sixths, *deciles* into tenths, etc. The first and ninth deciles (the 10th and 90th percentiles) are quite often used.

Cumulative Frequency

Percentiles and quartiles can easily be found from a cumulative frequency graph, as illustrated in the next example.

THE AVONFORD STAR

Letters to the editor

Dear Sir,

I am a student trying to live on my grant. My lecturers have given me a long list of textbooks to buy but if I get just half of them I will have nothing left to live on this month. The majority of the books on my list are over £16.

I want to do well at my studies but I won't do well without books and I won't do well if I am ill through not eating properly.

Please tell me what to do, and don't say to go to the library because the books I need are never there.

Yours sincerely,

Sheuli Roberts

After receiving this letter the editor wondered if there was a story in this. She asked a reporter to carry out a survey of the prices of textbooks in a large shop. The reporter took a random sample of 400 textbooks and the results are summarised in the table.

Cost, C (£)	Books
$C < 5$	11
$5 \leq C < 10$	47
$10 \leq C < 15$	85
$15 \leq C < 20$	133
$20 \leq C < 25$	79
$25 \leq C < 30$	36
$30 \leq C < 35$	9

Estimate the median, upper and lower quartiles of the cost of the books.

The first point to notice is that without the original data you cannot find the actual values required. All you can do is to make a reasonable estimate of where they might lie within their groups. A good way of doing this is to plot the cumulative frequency graph.

The first step is to make a cumulative frequency table:

Cost, C (£)	Frequency	Cost	Cumulative frequency
$0 \leq C < 5$	11	$C < 5$	11
$5 \leq C < 10$	47	$C < 10$	58
$10 \leq C < 15$	85	$C < 15$	143
$15 \leq C < 20$	133	$C < 20$	276
$20 \leq C < 25$	79	$C < 25$	355
$25 \leq C < 30$	36	$C < 30$	391
$30 \leq C < 35$	9	$C < 35$	400

NOTE

Check that you understand how the figures in the last column were calculated. Notice that the interval $C < 10$ means $0 \leq C < 10$ and so includes the 11 books in the interval $0 \leq C < 5$ and the 47 in the interval $5 \leq C < 10$, giving 58 in all. Similarly, to find the total for the interval $C < 15$, you must also add in the 85 books that lie within $10 \leq C < 15$, giving you $58 + 85 = 143$.

These figures are then graphed as a smooth curve. The median is read as number 200 (half of 400) and the quartiles as numbers 100 (¼ of 400) and 300 (¾ of 400), figure 1.10.

From the graph you can estimate that the upper quartile is £21.30, the lower quartile £13.20 and the median £17.20. The first and ninth deciles have also been marked on the graph.

Notice the distinctive shape of the cumulative frequency graph, sometimes called an *ogive*. It is like a letter S leaning forwards.

Sheuli said that more than half the books on her list were over £16. From the information on the cumulative frequency graph, only 165 out of the sample of 400 books (41¼%) cost under £16. So nearly 60% of books were over £16, backing up Sheuli's claim.

It could be helpful to draw a box-and-whisker plot but you do not know the greatest and least values. In this situation you draw the whiskers out to the first and ninth deciles (£8.60 and £25.50) and put arrows on the ends to show that they still have further to go, see figure 1.11.

NOTES

1. *You should be cautious about any conclusions you draw. This example deals with books and many books have prices like £9.95 and £39.99. In using a cumulative frequency curve you are assuming a reasonably continuous spread of the data throughout the intervals, and this may not always be the case.*
2. *You should also appreciate that the books that Sheuli needs may not be typical of all textbooks (and so presumably of those in this sample). Medical books, for example, tend to be very expensive.*

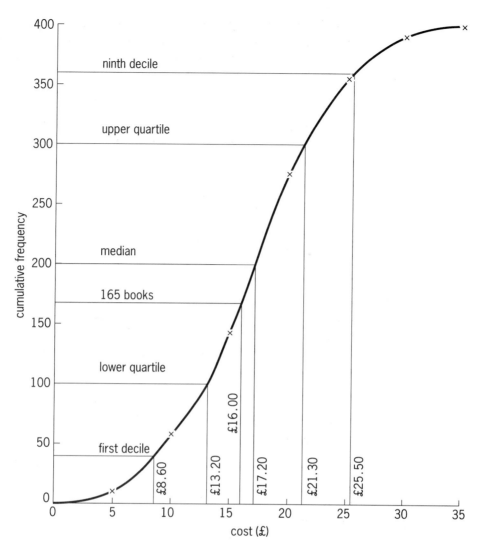

Figure 1.10

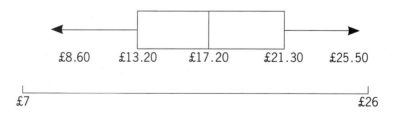

Figure 1.11

For Discussion

What other data do you think the newspaper editor would need to know before deciding whether to follow up the letter with an article about the financial problems of students; and where could the reporter find it?

Exercise 1D

1. For each of the following sets of figures, find (i) the range, (ii) the median, (iii) the upper and lower quartiles, (iv) the semi-interquartile range.

(a) 6   8   3   2   1   5   4   6   8   5   6   7   8   8   6

(b) 12   5   17   11   4   10   12   19   12   5   9   15   11   16   8
     18   12   8   9   11   12   14   8   14   7

(c) 25   28   29   30   20   23   23   27   25   28

(d) 115 123 132 109 127 116 128 132 114 109
     125 134 121 117 118 117 116 123 105 125

2. For each of the following sets of figures, find (i) the range, (ii) the median, (iii) the upper and lower quartiles, (iv) the semi-interquartile range.

(a)   2   8   4   6   3   5   1   8   2   5   8   0   3   7   8

(b) 12   18   14   16   13   15   11   18   12   15   18   10   13   17   18

(c) 272 278 274 276 273 275 271 278 272 275 278 270 273 277 278 2

(d) 20   80   40   60   30   50   10   80   20   50   80   0   30   70   80

3. Find (i) the median, (ii) the upper and lower quartiles and (iii) the semi-interquartile range for the scores of golfers in the first round of a competition.

70	I
71	II
72	IIII
73	Ⅼᚻᛏ III
74	Ⅼᚻᛏ Ⅼᚻᛏ II
75	Ⅼᚻᛏ II
76	Ⅼᚻᛏ
77	Ⅼᚻᛏ I
78	
79	III
80	I
81	
82	I

(iv) Illustrate the data with a box-and-whisker plot.

(v) The scores for the second round are illustrated on the box-and-whisker plot. Compare the two and say why you think the differences might have arisen.

67   68   70   74   77

4. The number of goals scored by a hockey team in its matches one season are illustrated on the vertical line chart.
   (i) Draw a box-and-whisker plot to illustrate the same data.
   (ii) State, with reasons, which you think is the better method of display in this case.

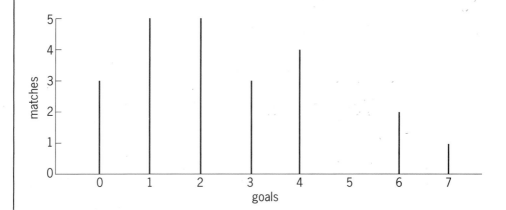

5. One year the yields, *y*, of a number of walnut trees were recorded in kg as follows:

Yield (kg)	Frequency
$40 \leq y < 50$	1
$50 \leq y < 60$	5
$60 \leq y < 70$	7
$70 \leq y < 80$	4
$80 \leq y < 90$	2
$90 \leq y < 100$	1

(i) Draw up the cumulative frequency table for these data.
(ii) Draw the cumulative frequency graph.
(iii) Use your graph to estimate the median and interquartile range of the yields.
(iv) Draw a box-and-whisker plot to illustrate the data.

The piece of paper where the actual figures had been recorded was then found, and these were:

44 59 67 76 52   62 68 78 53 63   69 82 53 65 70
85 93 56 65 74

(v) Use these data to find the median and interquartile range and compare your answers with those you obtained from the grouped data.
(vi) What are the advantages and disadvantages in grouping data?

Exercise 1D continued

6. The intervals of time between successive emissions from a weak radioactive source were measured for 200 consecutive intervals, with the following results:

Interval (seconds)	0–5	5–10	10–15	15–20	20–25	25–30	30–35
Frequency	23	67	42	26	21	15	6

(i) Draw a cumulative frequency curve for this distribution.
(ii) Use your graph to estimate (*a*) the median, (*b*) the interquartile range.
(iii) Calculate an estimate of the mean of the distribution.

[MEI]

7. In a sample of 800 eggs from an egg farm each egg was weighed and classified according to its mass. The frequency distribution was as follows:

Mass in grams	40–45	45–50	50–55	55–60	60–65	65–70
Number of eggs	36	142	286	238	76	22

Draw a cumulative frequency graph of these data, using a scale of 2 cm to represent 5 grams on the horizontal axis (which should be labelled from 40 to 70 grams) and a scale of 2 cm to represent 100 eggs on the vertical axis.

Use your graph to estimate for this sample:

(a) the percentage of eggs which would be classified as large (over 58 grams);
(b) the median mass of an egg;
(c) the interquartile range.

Indicate clearly on your diagram how you arrive at your results.

[MEI]

8. The table below shows the average productivity of coal miners in Great Britain over a 12-month period, measured in tonnes per manshift.

Productivity	Miners (thousands)	Productivity	Miners (thousands)
0–1	14	2.5–3.0	44
1–1.5	31	3.0–3.5	11
1.5–2.0	51	3.5–5.0	20
2.0–2.5	62		

(a) Illustrate this data by drawing a histogram.

(b) Find, in thousands, the total number of miners.

(c) Draw a cumulative frequency graph on a separate sheet.

(d) Estimate, from your graph, the median productivity.

(e) Estimate the total coal produced per shift.

[O & C]

9. The following table gives the age distribution of 3137 holders of a specific type of life assurance policy. (The true age of a policyholder whose nearest age is, for example, 44 lies between 43½ and 44½.)

Nearest age	Number	Nearest age	Number
21–25	44	61–65	396
26–30	47	66–70	488
31–35	66	71–75	471
36–40	107	76–80	358
41–45	147	81–85	222
46–50	151	86–90	104
51–55	208	91–95	18
56–60	305	96–100	5

Estimate the mean age of a policyholder.

Calculate also the median age and the upper and lower quartiles.

Plot the data on a cumulative frequency diagram and from this estimate the median age and the upper and lower quartiles and compare with the values previously calculated.

[MEI]

10. At a plant breeding institute, two different strains of a certain species of plant were compared by measuring the length, to the nearest mm, of each of 100 leaves of each strain. The results recorded were as follows.

Length of leaf (mm)	Frequency		Length of leaf (mm)	Frequency	
	Strain A	Strain B		Strain A	Strain B
Under 10	3	1	30–34	11	25
10–14	6	4	35–39	6	20
15–19	11	6	40–44	4	11
20–24	22	10	45–49	1	4
25–29	35	16	Over 50	1	3

Exercise 1D continued

Draw up a cumulative frequency table for each set of results. On the same axes draw the corresponding cumulative frequency graphs. (Scales: $2\,cm \equiv 10\,mm$ horizontally, $2\,cm \equiv 10$ leaves vertically.) Use your graphs to deduce which strain should be developed

(a) to produce plants with longer leaves;

(b) to produce plants with uniformity of length of leaf.

Give numerical readings taken from your graphs to support your conclusions.

A leaf is selected at random from a plant of each strain. What is the probability that the lengths of both leaves lie between 25 mm and 35 mm?

[MEI]

11. *Gross weekly earnings of full-time adults, April 1990 (£)*

| | Manual | | Non-manual | |
	Men	Women	Men	Women
Upper quartile	280	171	414	264
Median	221	137	312	191
Lower quartile	174	112	231	147

(Source: *New Earnings Survey*, 1990)

(i) Look at these figures and state what you conclude from them.

(ii) Using the semi-interquartile range as the measure, compare the spread of the earnings of women with that of men, both manual and non-manual.

(iii) A measure of spread relative to the size of the figures in a distribution is called an *absolute measure of dispersion*. There are several absolute measures; one of them, the *Quartile coefficient of dispersion*, is given by $(Q_3 - Q_1) \div (Q_3 + Q_1)$. Find this coefficient for each of the four categories (men/women manual/non-manual), and comment briefly on your results.

Variance and Standard Deviation

The most widely used measure of spread is *standard deviation*.

Although standard deviation is a function key on most scientific calculators, you should still know how to work it out on paper since you will then understand better what it actually means. The following example is worked with a tiny data set, just 4 items, purely to show the routines involved.

Suppose that you want to find the standard deviation of the numbers 4, 6, 5, 9.

First find the mean, $\bar{x}$, of the numbers.

$$\bar{x} = \frac{(4+6+5+9)}{4} = 6$$

Then find the *deviation* of each number, that is its difference from the mean.

x	deviation, $d\ (= x - \bar{x})$
4	$4 - 6 = -2$
6	$6 - 6 = 0$
5	$5 - 6 = -1$
9	$9 - 6 = 3$
	$\Sigma d = 0$

This symbol is the Greek upper case *sigma*, and means 'the sum of'

Notice that the sum of the deviations is zero. This should always be the case because the mean has been worked out so that the totals above and below it are the same. It provides you with a useful check; if your Σd does not come to zero you have made a mistake already!

If you ignore the minus signs of all the negative values of the deviation, you have the *modulus* or *absolute value* of d and your table now looks like this:

x	$d = x - \bar{x}$	$|d|$
4	$4 - 6 = -2$	2
6	$6 - 6 = 0$	0
5	$5 - 6 = -1$	1
9	$9 - 6 = 3$	3
	$\Sigma|d| =$	6

$|d|$ is the *modulus* of d and means the *positive value of d.*

You can now find the average of the positive values of the deviation, the *Mean Absolute Deviation (MAD)*:

$$\text{Mean Absolute Deviation} = \frac{6}{4} = 1.5.$$

Although it is a possible measure of dispersion, the Mean Absolute Deviation is not used much in Statistics, just in a few specialised applications.

Another way of dealing with the negative deviations is to square the d values. The mean of these squared d values, the *Mean Squared Deviation (MSD)*, is called the *variance;* the square root of the variance is the *standard deviation*, s.

x	$d = x - \bar{x}$	d^2
4	$4 - 6 = -2$	4
6	$6 - 6 = 0$	0
5	$5 - 6 = -1$	1
9	$9 - 6 = 3$	9
	$\Sigma d^2 =$	14

$$\text{MSD} = \text{Variance}, \ s^2 = \frac{\Sigma d^2}{n} = 14 \div 4 = 3.5$$

Standard deviation, $s = \sqrt{3.5} = 1.87$

This procedure can be expressed by the formula:

$$s = \sqrt{\frac{\Sigma(x - \bar{x})^2}{n}}$$

An alternative procedure is given by the formula:

$$s = \sqrt{\frac{\Sigma x^2}{n} - \bar{x}^2}$$

When you apply the second formula to the same set of numbers, 4, 6, 5 and 9, you get the same answer.

x	x^2
4	16
6	36
5	25
9	81
$\Sigma \ 24$	158

$$\text{Variance}, \ s^2 = \frac{158}{4} - \left(\frac{24}{4}\right)^2$$

Standard deviation, $s = \sqrt{3.5} = 1.87$

The answers are the same as before.

The second formula can be easily remembered in this form:

Variance = the mean of the squares minus the square of the mean.

The proof that these two formulae are equivalent is as follows.

The first formula is

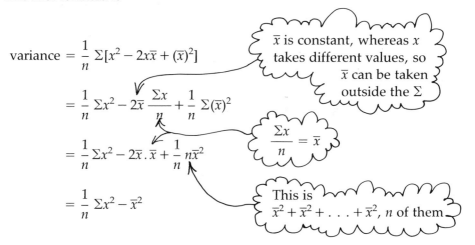

$$\text{variance} = \frac{1}{n}\Sigma[x^2 - 2x\bar{x} + (\bar{x})^2]$$

$\bar{x}$ is constant, whereas x takes different values, so $\bar{x}$ can be taken outside the Σ

$$= \frac{1}{n}\Sigma x^2 - 2\bar{x}\frac{\Sigma x}{n} + \frac{1}{n}\Sigma(\bar{x})^2$$

$$\frac{\Sigma x}{n} = \bar{x}$$

$$= \frac{1}{n}\Sigma x^2 - 2\bar{x}.\bar{x} + \frac{1}{n}n\bar{x}^2$$

$$= \frac{1}{n}\Sigma x^2 - \bar{x}^2$$

This is $\bar{x}^2 + \bar{x}^2 + \ldots + \bar{x}^2$, n of them

which is the second formula.

NOTE

In more advanced work, it is more usual to replace n by n − 1 in the formula

$$s^2 = \frac{\Sigma(x - \bar{x})^2}{n}$$

(having used n to calculate $\bar{x}$).

The reason for this is that although there are n data items, the effect of using them to find the mean is that you have only n − 1 independent deviations when you then go on to find the variance and standard deviation.

The value of variance found in this way is also denoted by $\hat{\sigma}^2$.

Many calculators allow you the option of using n or n − 1 when finding standard deviation, with keys marked σ_n or σ_{n-1}.

The calculation of variance and standard deviation would normally be carried out in a table (or within a spreadsheet).

If the data are sorted so that x_i occurs with frequency f_i, then these formulae are written:

$$\bar{x} = \frac{1}{n}\sum_i f_i x_i \qquad \text{where } \sum_i f_i = n$$

$$s^2 = \frac{1}{n}\sum_i f_i(x_i - \bar{x})^2 = \frac{1}{n}\sum_i f_i x_i^2 - \bar{x}^2$$

Standard Deviation of Grouped Data

To estimate the standard deviation of grouped data, you first assume that all the values in each interval are equally spaced around its midpoint. The calculation is shown in the example below using the data on the prices of textbooks from p. 33.

Cost, C (£)	Mid interval x	Frequency f	fx	x^2	fx^2
$0 \leq C < 5$	2.5	11	27.5	6.25	68.75
$5 \leq C < 10$	7.5	47	352.5	56.25	2643.75
$10 \leq C < 15$	12.5	85	1062.5	156.25	13281.25
$15 \leq C < 20$	17.5	133	2327.5	306.25	40731.25
$20 \leq C < 25$	22.5	79	1777.5	506.25	39993.75
$25 \leq C < 30$	27.5	36	990	756.25	27225
$30 \leq C < 35$	32.5	9	292.5	1056.25	9506.25
Σ		400 n	6830 Total		133450

$$\text{Mean, } \bar{x} = \frac{6830}{400} = 17.075$$

$$\text{Variance, } s^2 = \frac{133\,450}{400} - 17.075^2 = 42.069375$$

$$\text{Standard deviation, } s = 6.486091$$

Since this is only an estimate (because it is calculated from grouped data), it would be quite reasonable to give the value of the standard deviation as £6.5.

Coding

Calculations of mean and standard deviation can often be simplified a great deal by *coding*. In the example above, this involves replacing x by y where the values of y are found by

1. Taking 2.5 away from all the values of x
2. Dividing the remaining value by 5 to get y.

In this case $x = 32.5$ is coded into $y = \dfrac{(32.5 - 2.5)}{5} = 6$,

and the coding can be written algebraically as

$$y = \frac{x - 2.5}{5} \quad \text{or} \quad x = 5y + 2.5$$

You carry out the calculations with these simplified numbers and then decode your answers.

| Cost, C (£) | Midinterval | Frequency | | | | |
	x	y	f	fy	y^2	fy^2
$0 \le C < 5$	2.5	0	11	0	0	0
$5 \le C < 10$	7.5	1	47	47	1	47
$10 \le C < 15$	12.5	2	85	170	4	340
$15 \le C < 20$	17.5	3	133	399	9	1197
$20 \le C < 25$	22.5	4	79	316	16	1264
$25 \le C < 30$	27.5	5	36	180	25	900
$30 \le C < 35$	32.5	6	9	54	36	324
			Σ 400	1166		4072
			n	Total		

Coded Mean, $\bar{y} = \dfrac{1166}{400} = 2.915$

Variance, $s_y^2 = \dfrac{4072}{400} - 2.915^2 = 1.682775$

Standard deviation, $s_y = \sqrt{1.682775} = 1.2972182$

The answers now have to be decoded by reversing the coding steps

Uncoded Mean, $\bar{x} = 2.915 \times 5 + 2.5 = 17.075$

Standard deviation, $s_x = 1.2972182 \times 5 = 6.486091$

Variance, $s_x^2 = 1.682775 \times 5^2 = 42.069376$

Notice that standard deviation and variance both measure spread and this is not altered by shifting all the numbers by a fixed amount. Consequently there is no need to compensate for reducing all the numbers by 2.5, whereas this does have to be added back on to the mean.

The decoding process can be written generally as:

If $\qquad\qquad\qquad\qquad x = ay + b$

then

Mean: $\qquad\qquad\qquad \bar{x} = a\bar{y} + b$

Variance: $\qquad\qquad\qquad s_x^2 = a^2 s_y^2$

Standard deviation: $\qquad\quad s_x = |a| s_y$

N O T E *Notice the use of $|a|$ when decoding standard deviation. The value of standard deviation is always positive (or zero) so even if a negative value of a is used for coding, the final value, s_x, must be positive.*

Outliers

For most data sets:
about ⅔ of the values lie within 1 standard deviation of the mean,
about 95% lie within 2 standard deviations of the mean, and
about 99.5% within 3 standard deviations.

A simple test for identifying outliers is that they are those values which
lie more than 2 standard deviations away from the mean. This should be
seen not so much as a means of defining outliers, however, as a working
check on which data items it might be a good idea to look at carefully. It
is designed to pick out the 5% of values furthest from the mean.

For Discussion

1. On a standard IQ scale, the mean is 100 and the standard deviation is
   15.
   (i) Between what limits do the IQs of the central ⅔ of the population
        lie?
   (ii) What proportion of the population have IQs
        (a) above 130, (b) below 85, (c) above 145?

2. British adult males have a mean height of about 5′8″. By considering
   the range of heights for the central 95% of adult males, *estimate* the
   standard deviation.

3. In an examination 10% of candidates score 75 or more and get a grade
   A, and 20% of candidates score below 40 and get a grade F. *Estimate*
   the mean and standard deviation of the marks.

Exercise 1E

1. For each of the following sets of figures, find (i) the mean (ii) the
   standard deviation.

   (a)   6   8   3   2   1   5   4   6   8   5
         6   7   8   8   6

   (b)   12   5   17   11   4   10   12   19   12   5
         9   15   11   16   8
         18   12   8   9   11   12   14   8   14   7

   (c)   25   28   29   30   20   23   23   27   25   28

   (d)   115 123 132 109 127   116 128 132 114 109
         125 134 121 117 118   117 116 123 105 125

   These figures are the same as those in question 1 of exercise 1D.
   (iii) Compare your answers with those for the median and semi-
         interquartile range. (Answers for exercise 1D are on p. 136.)

2. For each of the following sets of figures, find (i) the mean (ii) the standard deviation.

   (a)   2   8   4   6   3   5   1   8   2   5
         8   0   3   7   8

   (b) 12 18 14 16 13   15 11 18 12 15
        18 10 13 17 18

   (c) 272 278 274 276 273   275 271 278 272 275
        278 270 273 277 278

   (d) 20 80 40 60 30   50 10 80 20 50
        80   0 30 70 80

These figures are the same as those in question 2 of exercise 1D.
(iii) Compare your answers with those for the median and semi-interquartile range.

3. The scores of 20 rounders clubs one Saturday were as follows:

Rounders	0–1	1½–2½	3–6	6½–9½	10–20
Clubs	3	2	4	5	6

Calculate the mean and standard deviation of the number of rounders scored per club and comment on the accuracy of your answers.

4. The graph shows the number of kittens weaned per kindle by a sample of female cats. Only one kindle of kittens was considered per mother cat.

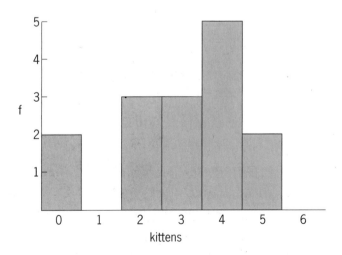

   (i)   Calculate the mean and standard deviation of this data.
   (ii) If the mean kindle size at birth was 3.8, what percentage of kittens survived to weaning?

5. Calculate the mean and the standard deviation of the numbers of letters in the words of this sentence.

[SMP]

6. A manufacturer produces electrical cable which is sold on reels. The reels are supposed to hold 100 metres of cable. In the quality control department the length of cable on randomly chosen reels is measured. These measurements are recorded as deviations, in cm, from 100 m. (So, for example, a length of 99.84 m is recorded as -16.)

For a sample of 20 reels the recorded values, x, are summarised by

$$\Sigma x = -86 \qquad \Sigma x^2 = 4281$$

(i) Calculate the mean and standard deviation of the values of x.
(ii) Hence find the mean and standard deviation, in metres, of the lengths of cable on the 20 reels.
(iii) Later it is noticed that one of the values of x is -47, and it is thought that so large a value is likely to be an error. Give a reason to support this view.
(iv) Find the new mean and standard deviation of the values of x when the value -47 is discarded.

[MEI]

7. The table summarises the observed lifetimes x, in seconds, of 50 fruit flies subjected to a new spray in a controlled experiment.

Interval	Mid-interval value	Frequency
$0.5 \le x < 5.5$	3	3
$5.5 \le x < 10.5$	8	22
$10.5 \le x < 15.5$	13	12
$15.5 \le x < 20.5$	18	9
$20.5 \le x < 25.5$	23	2
$25.5 \le x < 30.5$	28	1
$30.5 \le x < 35.5$	33	1

(i) Making clear your methods and showing all your working, estimate the mean m and the standard deviation s of these lifetimes. Give your answers correct to 3 significant figures and do not make any corrections for grouping.
(ii) Draw the cumulative frequency curve, and use your graph to estimate the minimum lifetime below which 70% of all lifetimes lie.

[MEI]

8. The table shows the numbers of club ranking points scored by a
bridge player in a series of 30 games.

6 6 3 8 7 9 8 6 5 0 8 7 6 9 8
6 7 2 8 7 6 6 8 9 7 4 8 7 5 9

Use your calculator to find the mean and standard deviation of these
scores.

In a further 20 games, data on the player's scores were summarised
by $\Sigma x = 143$, $\Sigma x^2 = 1071$. Find the mean and standard deviation of the
scores on all 50 games.

[MEI]

9. As part of a Biology experiment Andrew caught and weighed 120
minnows. He used his calculator to find the mean and standard
deviation of their weights:

$$\text{Mean} \quad \bar{x} = 26.231 \text{ g}$$
$$\text{SD} \quad s_x = 4.023 \text{ g}$$

(i) Find the total weight, Σx, of Andrew's 120 minnows

(ii) Use the formula $s_x = \sqrt{\dfrac{\Sigma x^2}{n} - \bar{x}^2}$

to find Σx^2 for Andrew's minnows.

Another member of the class, Sharon, did the same experiment with
minnows caught from a different stream. Her results are summarised
by:

$$n = 80$$
$$\bar{x} = 25.214$$
$$s_x = 3.841$$

Their teacher says they should combine their results into a single set
but they have both thrown away their measurements.
(iii) Find n, Σx and Σx^2 for the combined data set.
(iv) Find the mean and standard deviation for the combined data set.

10. A university admissions officer interviews 30 candidates on Monday
and 20 on Tuesday. On each day the mean length of all the interviews
is exactly 15 minutes, but the standard deviations are 2 minutes on
Monday and 2½ minutes on Tuesday. Find the sums of the squares
of the deviations from the mean on each day, and hence find (correct
to 1 decimal place) the standard deviation for the two days taken
together.

[SMP]

11. A random sample of 200 batteries, of nominal potential 6 V, was taken from a very large batch of batteries. The potential difference between the terminals of each battery was measured, resulting in the following table of data:

Potential difference in volts (mid-interval value)	Number of batteries
5.80	1
5.85	4
5.90	22
5.95	42
6.00	60
6.05	44
6.10	24
6.15	2
6.20	1

Calculate the mean and standard deviation of these voltages and illustrate the data on a histogram. Mark clearly on the histogram the mean voltage and the voltages which are 2 standard deviations either side of the mean.

[MEI]

12. A large manufacturing firm orders a component from two suppliers, *Neverlate Ltd* and *Ontime Ltd*. The delivery times to the nearest day for two random samples of 10 orders, one from each supplier, were as follows:

Neverlate Ltd	10 10 9 10 11 11 9 11 10 10
Ontime Ltd	8 10 14 7 10 11 10 8 15 12

(i)   Illustrate the data on two histograms, one for each supplier. (Put your histograms side by side, or one under the other, so that you can compare them easily.)

(ii) For each supplier, calculate, from the above data, estimates of the mean and standard deviation of the delivery times to 3 significant figures.

(iii) The firm has decided to use only one supplier for the component in future. Assume that you are in charge of ordering. Write a concise memorandum for your Managing Director recommending one supplier and giving clear reasons for your choice based on the results in (i) and (ii) above.

[MEI]

13. The data below record, in ranked order, the deaths per 100 000 population from motor vehicle accidents in the 50 states of the USA in 1973.

13.5 16.3 16.8 16.9 17.4 18.3 18.4 18.7 20.2 21.1
21.8 22.4 22.6 23.0 23.8 24.0 24.5 25.3 25.4 26.3
26.6 27.3 27.6 28.0 28.1 28.5 29.9 30.2 30.4 30.8
31.2 31.3 32.5 33.0 33.2 33.4 34.6 34.6 34.9 35.5
35.8 38.7 39.8 41.8 44.8 45.3 46.6 48.7 54.4 57.7

In parts (i) and (ii) of this question you will need to use the fact that the data are obtained from the true figures by rounding correct to 1 decimal place. For example the first entry of 13.5 indicates a true figure between 13.45 and 13.55.

(i) Summarise these data in a frequency distribution table so that, if x denotes a data point, the intervals are $9.95 \leq x < 14.95$, $14.95 \leq x < 19.95, \ldots, 54.95 \leq x < 59.95$. Plot the data on a histogram.

(ii) From the histogram, estimate the 10th percentile, i.e. the lowest death rate for which 10% of the states have death rates at or below that rate. Compare your answer with the correct value obtained directly from the table.

(iii) Given that $\Sigma x = 1491.9$ and $\Sigma x^2 = 49\,401.05$, calculate the mean and standard deviation. Find also the median death rate by using the table.

(iv) On your histogram, indicate the positions of the mean and median. Comment very briefly on the shape of the histogram.

[MEI]

14. The weights of 225 fifteen-year-old children were measured, and the following frequency table was constructed.

Mid-interval weight (lb) = w	Number = f	Mid-interval weight (lb) = w	Number = f
75	1	115	30
80	3	120	31
85	6	125	29
90	9	130	21
95	13	135	11
100	17	140	4
105	22	145	2
110	26		

$\Sigma f = 225$, $\Sigma fw = 25\,600$, $\Sigma fw^2 = 2\,957\,650$.

Illustrate the data in a histogram.
Estimate (to 1 decimal place) the mean, median and standard deviation of these weights, using (where necessary) the results at the end of the table. (Contd.)

Exercise 1E continued

Frequency distributions which are not symmetrical about their mean are called skew; the parameter S, where

$$S = \frac{(\text{mean} - \text{median})}{\text{standard deviation}},$$

is sometimes used as a measure of the degree of skewness. Estimate S for the data given in the table. What can be deduced about the shape of another histogram for which S has the same magnitude as that for the data above, but the opposite sign?

Give one reason why S is a better measure of the degree of skewness than S', where $S' = (\text{mean} - \text{median})$.

[MEI]

KEY POINTS

In this chapter you have met various techniques for investigating raw data. When you are doing so, you may use some or perhaps all of these, but when you come to summarise your findings you must be concise.

Giving pages of figures and diagrams may show that you have done a lot of work, but that is not the object of the exercise when you are exploring raw data. Your aim must be to find out and summarise the main features of the information that is locked up in the figures, and to give some interpretation of them.

The purpose of computation is insight not numbers.

Hamming

Exercise 1F

1. The full details of the 32 cycling accidents reported in *The Avonford Star* are given opposite. Investigate these and the data from the survey of cyclists' ages (given below) and write a report on what you find.

Ages of cyclists (from survey)

66	6	62	19	20		15	21	8	21	63		44	10	44	34	18
35	26	61	13	61		28	21	7	10	52		13	52	20	17	26
64	11	39	22	9		13	9	17	64	32		8	9	31	19	22
37	18	138	16	67		45	10	55	14	66		67	14	62	28	36
9	23	12	9	37		7	36	9	88	46		12	59	61	22	49
18	20	11	25	7		42	29	6	60	60		16	50	16	34	14
18	15														

Name	Age	Distance from home	Cause	Injuries	Treatment
John Smith	45	3 km	skid	Concussion	Hosp. Outpatient
Debbie Lane	5	75 m	hit kerb	Broken arm	Hosp. Outpatient
Arvinder Sethi	12	1200 m	lorry	Multiple	Hosp. 3 weeks
Marion Wren	8	300 m	hit each other	Bruising	Hosp. Outpatient
David Huker	8	50 m		Concussion	Hosp. overnight
Paul Edwards	75	200 m	car	Sprained leg	Local Doctor
Colette Chesson	21	4.5 km	car	Multiple	Hosp. 2 weeks
Millie Smith	88	35 m	bollard	Shock	Local Doctor
Claude Benett	11	60 m	ditch	Broken arm	Hosp. Outpatient
Brendon Stanton	47	120 m	bicycle	Cuts & bruises	Mum a nurse
Ann Nelson	18	1.25 km	hit kerb	Hand injury	Local Doctor
Sara Moraga	13	3.5 km	car door	Bruises	—
Ludmila Steyn	13	30 m	lorry	Concussion	Hosp. overnight
Ken Devereaux	9	75 m	brake failure	Concussion	Hosp. overnight
Alice Walusimbi	62	250 m	car	Shock	Outpatient
Oliver Hill	12	15 m	lorry	Multiple	Hosp. 1 month
Rex Fox	61	4 km	dog	Minor	—
Thomas Izod	9	10 m	bicycle	Sprained wrist	Neighbour
Evelyn Sertas	15	150 m	car	Broken leg	Hosp. overnight
Margaret Rex	25	1.5 km	car	Collar bone	Hosp. Outpatient
Helen Smith	14	250 m	hit kerb	Cuts	—
Ping Chiu Liang	64	25 m	car	Internal	Hosp. 2 months
Agnes De Pellette	68	400 m	child	Concussion	Hosp. overnight
John Pothitos	28	2400 m	car	Bruises	—
Eric Snape	65	140 m	kerb	Cuts & bruises	—
Louise Teboul	65	45 m	bollard	Broken wrist	Hosp. Outpatient
Ahchun Fu	9	5 m	car	Broken leg	Hosp. Outpatient
Shangara Madar	35	1.2 km	car	Minor	—
Peter Owen	8	0.25 km	tree	Cuts	—
Denise Cropper	7	50 m	van	Concussion	Hosp. Outpatient
Peter Bland	66	250 m	car	Sprained wrist	—
Sue Parish	37	450 m	van	Broken arm	Hosp. Outpatient

2. Collect a substantial set of raw data (minimum size 50 items) about a topic that interests you and describe and interpret it in a written report.

2

Probability

If we knew Lady Luck better, Las Vegas would still be a roadstop in the desert.

Stephen Jay Gould

A library without books

If you plan to pop into the Avonford library and pick up the latest best seller, then forget it. All the best books 'disappear' practically as soon as they appear on the shelves.

I talked about the problem with the local senior librarian, Gina Clarke.

"We have a real problem with unauthorised loans at the moment," Gina told me. "Out of our total stock of, say, 80 000 books, something like 44 000 are out on loan at any one time. About 20 000 are on the shelves and I'm afraid the rest are unaccounted for."

That means that the probability of finding the particular book you want is exactly 1/4. With odds like that, don't bet on being lucky next time you visit your library.

Librarian Gina Clarke is worried about the problem of 'disappearing books'

How do you think the figure of ¼ at the end of the article was arrived at? Do you agree that the probability is *exactly* ¼?

The information about the different categories of book can be summarised as follows.

Category of books	Typical numbers
On the shelves	20 000
Out on loan	44 000
Unauthorised loan	16 000
Total stock	80 000

On the basis of these figures, it is possible to estimate the probability of finding the book you want. Of the total stock of 80 000 books bought by the library, you might expect to find about 20 000 on the shelves at any one time. As a fraction, this is $^{20}/_{80}$ or $\frac{1}{4}$ of the total. So, as a rough estimate, the probability of your finding a particular book is 0.25 or 25%.

Similarly, 16 000 out of the total of 80 000 books are on unauthorised loan — a euphemism for *stolen* — and this is 20%, or $\frac{1}{5}$.

An important assumption underlying these calculations is that all the books are equally likely to be unavailable, which is not very realistic since popular books are more likely to be stolen. Also, the numbers given are only rough approximations, so it is definitely incorrect to say that the probability is *exactly* $\frac{1}{4}$.

Measuring Probability

Probability (or chance) is a way of describing the likelihood of different possible outcomes or *events* occurring as a result of some *trial*.

In the example of the library books, the trial is looking in the library for a particular book. Let us assume that you already know that the book you want is on the library's stocks. The three possible outcomes (or events) are that the book may be: on the shelves; out on loan; or missing.

It is important in probability to distinguish trials from the events which they may generate. Here are a few examples.

Trials	Possible outcomes (events)
• guessing the answer to a four-option multiple choice question	A B C D

Trials	Possible outcomes (events)
• predicting the stamp on the next letter I receive	first class second class foreign other
• tossing a coin	heads tails

Estimating Probability

Probability is a number which measures likelihood. It may be estimated experimentally or theoretically.

Experimental Estimation of Probability

In many situations probabilities are estimated on the basis of data collected experimentally, as in the following example.

Of 30 drawing pins tossed in the air, 21 of them were found to have landed with their pins pointing up. From this you would estimate the probability that the next pin tossed in the air will land with its point up to be $^{21}/_{30}$ or 0.7.

You can describe this in more formal notation.

number of times it landed pin-up

$$P(U) = \frac{n(U)}{n(T)}$$

probability of the next throw landing pin-up

total number of throws

Theoretical Estimation of Probability

There are however some situations where you do not need to collect data to make an estimate of probability.

For example, when tossing a coin, common sense tells you that there are only two realistic outcomes and, given the symmetry of the coin, you would expect them to be equally likely. So the probability, $P(H)$, that the next coin will produce the outcome Heads can be written as follows:

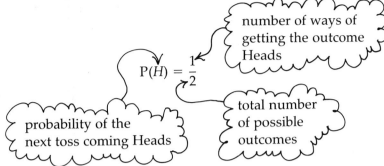

EXAMPLE   Using the notation described above, write down the probability that the correct answer for the next four-option multiple choice question will be answer A. What assumptions are you making here?

Solution
Provided that the test-setter has used each letter equally often, the probability, $P(A)$, that the next question will give answer A can be written as follows:

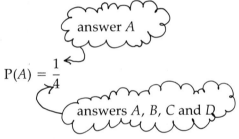

Notice that we have assumed that the four options are equally likely. Equiprobability is an important assumption underlying most work on probability.

Expressed formally, the probability, $P(A)$, of event A occurring is:

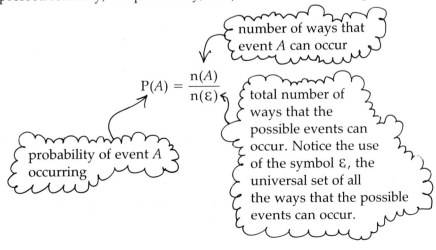

Probabilities of 0 and 1

The two extremes of probability are *certainty* at one end of the scale and *impossibility* at the other. Here are examples of certain and impossible events.

Trials	Certain events	Impossible events
tossing a single die	the result is in the range 1 to 6 inclusive	the result is a 7
tossing a coin	getting either heads or tails	getting neither heads nor tails

Certainty As you can see from the table above, for events that are certain, the number of ways that the event can occur, $n(A)$ in the formula, is equal to the total number of possible events, $n(\varepsilon)$.

$$\frac{n(A)}{n(\varepsilon)} = 1$$

So the probability of an event which is certain = 1.

Impossibility For impossible events, the number of ways that the event can occur, $n(A)$, is zero.

$$\frac{n(A)}{n(\varepsilon)} = \frac{0}{n(\varepsilon)} = 0$$

So the probability of an event which is impossible = 0.

Typical values of probabilities might be something like 0.3 or 0.9. If you arrive at probability values of, say, -0.4 or 1.7, you will know that you have made a mistake since these are meaningless.

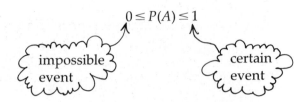

$$0 \le P(A) \le 1$$

impossible event   certain event

The Complement of an Event

The complement of an event A, denoted by A', is the event not-A, that is the event "A does not happen".

EXAMPLE It was found that out of a box of 50 matches, 45 lit but the others did not. What was the probability that a randomly selected match would not have lit?

Solution

The probability that a randomly selected match lit was

$$P(A) = 45/50 = 0.9$$

The probability that a randomly selected match did not light was

$$P(A') = (50 - 45)/50 = 5/50 = 0.1$$

From this example you can see that

$$P(A') = 1 - P(A)$$

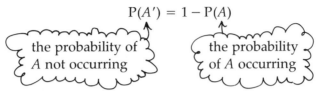

the probability of *A* not occurring   the probability of *A* occurring

This is illustrated in figure 2.1.

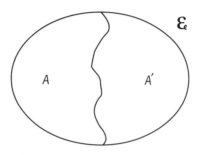

Figure 2.1 Venn diagram showing events *A* and Not-*A* (*A'*)

Expectation

How many people can the Health Area expect to contract 'flu'? The answer is easily seen to be $120\,000 \times \frac{1}{3} = 40\,000$. This is called the *expectation* and is given by np where n is the population size and p the probability.

Expectation is a technical term and need not be a whole number. Thus the expectation of the number of heads when a coin is tossed 5 times is $5 \times \frac{1}{2} = 2.5$. You would be wrong to go on and say "That means either 2 or 3" or to qualify your answer as "about 2½". The expectation is 2.5.

The Probability of Either One Event or Another

So far we have looked at just one event at a time. However, it is often useful to bracket two or more of the events together and calculate their combined probability.

EXAMPLE

The table below is based on the data at the beginning of this chapter and shows the probability of the next book requested falling into each of the three categories listed, assuming that each book is equally likely to be requested.

Category of books	Typical Numbers	Probability
On the shelves (S)	20 000	0.25
Out on loan (L)	44 000	0.55
Unauthorised loan (U)	16 000	0.20
Total ($S + L + U$)	80 000	1.00

What is the probability that a randomly requested book is *either* out on loan *or* on unauthorised loan (ie that it is not available).

Solution

No. of books on loan   No. of books on unauthorised loan

$$P(L \text{ or } U) = \frac{44\,000 + 16\,000}{80\,000}$$

No. of books

$$= \frac{60\,000}{80\,000} = 0.75$$

This can be written in more formal notation as

$$P(L \cup U) = \frac{n(L \cup U)}{n(\varepsilon)} = \frac{n(L)}{n(\varepsilon)} + \frac{n(U)}{n(\varepsilon)}$$

$$P(L \cup U) = P(L) + P(U)$$

Notice the use of the *union* symbol, $\cup$, to mean *or*.
This is illustrated in figure 2.2.

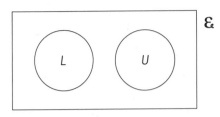

key: L = out on loan
U = out on unauthorised loan

Figure 2.2 Venn diagram showing events *L* and *U*. It is not possible for both to occur

In this example you could add the probabilities of the two events to get the combined probability of *either one or the other* event occurring.

However you have to be very careful adding probabilities as you will see in the next example.

EXAMPLE

Below are further details of the categories of books in the library.

Category of books	Number of books
on the shelves	20 000
out on loan	44 000
adult fiction	22 000
adult non-fiction	40 000
junior	18 000
unauthorised loan	16 000
Total stock	80 000

Assuming all the books in the library are equally likely to be requested, find the probability that the next book requested will be either out on loan or a book of adult non-fiction.

Solution

$$P(\text{loan}) + P(\text{adult non-fiction}) = \frac{44\,000}{80\,000} + \frac{40\,000}{80\,000} = 0.55 + 0.5$$

$$= 1.05$$

This is clearly nonsense. You cannot have a probability greater than 1.

So what has gone wrong?

The way this calculation was carried out involved some double counting. Some of the books classed as adult fiction were counted again because they were in the on-loan category, as you can see in figure 2.3.

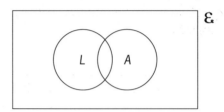

key: L = loaned
A = adult non–fiction

Figure 2.3 Venn diagram showing events L and A. It is possible for both to occur

If you add all six of the book categories together, you find that they add to 160 000, which represents twice the total number of books owned by the library.

A more useful representation of these data is given in the two-way table below.

	Adult fiction	Adult non-fiction	Junior	Total
Shelved	4000	12 000	4000	20 000
Loaned	**14 000**	20 000	**10 000**	**44 000**
Unauthorised loan	4000	8000	4000	16 000
TOTAL	22 000	**40 000**	18 000	80 000

If you simply add the 44 000 and the 40 000, you *double count* the 20 000 books which fall into both categories. So you need to subtract the 20 000 to ensure that it is counted only once. Thus:

number either out on loan or adult non-fiction

$$= 44\,000 + 40\,000 - 20\,000$$
$$64\,000 \text{ books}$$

So, the required probability $\quad = \dfrac{64\,000}{80\,000} = 0.8$

Mutually Exclusive Events

The problem of double counting does not occur when adding two rows in the table. Two rows cannot overlap, or *intersect*, which means that those categories are *mutually exclusive* (i.e. the one excludes the other). The same is true for two columns within the table.

Where two events, A and B, are mutually exclusive, the probability that either A or B occurs is equal to the sum of the separate probabilities of A and B occurring.

Where two events, A and B, are *not* mutually exclusive, the probability that either A or B occurs is equal to the sum of the separate probabilities of A and B occurring minus the probability of A and B occurring together.

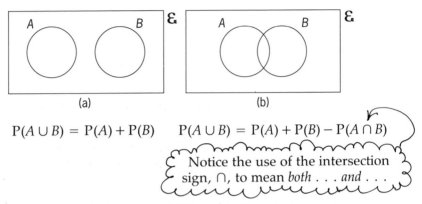

$$P(A \cup B) = P(A) + P(B) \qquad P(A \cup B) = P(A) + P(B) - P(A \cap B)$$

Notice the use of the intersection sign, $\cap$, to mean *both . . . and . . .*

Figure 2.4 (a) Mutually exclusive events; (b) Not mutually exclusive events

EXAMPLE

A fair die is thrown. What is the probability that it shows
(i)   Event A: an even number
(ii) Event B: a number greater than 4
(iii) Event $A \cup B$: a number which is either even or greater than 4?

Solution
(i)   Event A:
      3 out of the 6 numbers on a die are even, namely 2, 4 and 6.
      So $P(A) = \frac{3}{6} = \frac{1}{2}$.
(ii) Event B:
      2 out of the 6 numbers on a die are greater than 4, namely 5 and 6.
      So $P(B) = \frac{2}{6} = \frac{1}{3}$.
(iii) Event $A \cup B$:
      4 of the numbers on a die are either even or greater than 4, namely 2, 4, 5 and 6.
      So $P(A \cup B) = \frac{4}{6} = \frac{2}{3}$.
      This could also be found using

$$P(A \cup B) = P(A) + P(B) - P(A \cap B)$$
$$P(A \cup B) = \frac{3}{6} + \frac{2}{6} - \frac{1}{6}$$
$$= \frac{4}{6} = \frac{2}{3}$$

This is the number 6 which is both even and greater than 4.

Exercise 2A

1. Three separate electrical components – switch, bulb and contact point – are used together in the construction of a pocket torch. Of 534 defective torches, examined to identify the cause of failure, 468 are found to have a defective bulb. For a given failure of the torch, what is the probability that either the switch or the contact point is responsible for the failure? State clearly any assumptions that you have made in making this calculation.

2. If a fair die is thrown, what is the probability that it shows (i) 4; (ii) 4 or more; (iii) less than 4; (iv) an even number?

3. A bag containing Scrabble letters has the following letter distribution

A	B	C	D	E	F	G	H	I	J	K	L	M
9	2	2	4	12	2	3	2	9	1	1	4	2

N	O	P	Q	R	S	T	U	V	W	X	Y	Z
6	8	2	1	6	4	6	4	2	2	1	2	1

The first letter is chosen at random from the bag, find the probability that it is
(i) an E; (ii) in the first half of the alphabet; (iii) in the second half of the alphabet; (iv) a vowel; (v) a consonant; (vi) the only one of its kind.

4. A lottery offers five prizes, each of £100, and a total of 2000 lottery tickets are sold. You buy a single ticket for 20 pence.
   (i)   What is the probability that you will win a prize?
   (ii) What is the probability that you will not win a prize?
   (iii) How much money do the lottery organisers stand to make or lose?
   (iv) How much money should the lottery organisers charge for a single ticket in order to break even?
   (v)   If they continue to charge 20 pence per ticket, how many tickets would they need to sell in order to break even?

5. **A sporting chance.**
   (a) Two players, A and B, play tennis. On the basis of their previous results, the probability of A winning, $P(A)$, is calculated to be 0.65. What is $P(B)$, the probability of B winning?
   (b) Two hockey teams, A and B, play a game. On the basis of their previous results, the probability of team A winning, $P(A)$, is calculated to be 0.65. Why is it not possible to calculate directly $P(B)$, the probability of team B winning, without further information?
   (c) In a tennis tournament, player A, the favourite, is estimated to have a 0.3 chance of winning the competition. Player B is

estimated to have a 0.15 chance. Find the probability that either A or B will win the competition.

(d) In the Five Nations Rugby Championship, France and England are given a 25% chance of winning or sharing the championship cup. It is also estimated that there is a 10% chance that they will share the cup. What is the probability that either England or France will win or share the cup?

6. The integers from 1 to 20 are written on separate identical cards and the cards are then thoroughly shuffled. The cards can show numbers which are Even (E), Odd (O) or Squares (S).

A card is chosen at random. Find the probability that the number showing is:

(i)   Even, E; (ii) Square, S; (iii) Odd, O;
(iv) Both even and square, $E \cap S$;
(v)   Either even or square, $E \cup S$;
(vi) Both even and odd, $E \cap O$;
(vii) Either even or odd, $E \cup O$.

Write down equations connecting the probabilities of the following events:

(viii) $E, S, E \cap S, E \cup S$; (ix) $E, O, E \cap O, E \cup O$

7. The data in the table below show the numbers of part-time students in higher education in 1988/89 by sex and type of establishment (numbers in thousands).

	Women	Men
Universities	21.1	29.0
Open University	40.3	45.0
Polytechnics and colleges		
– part-time day courses	67.0	118.5
– evening-only courses	26.5	38.1
Total part-time students	154.9	230.6

(Source: *Education Statistics for the United Kingdom*, Department of Education and Science.)

Find the probability that a part-time student chosen at random is:
(i)   An Open University student
(ii) Female
(iii) A female Open University student
(iv) Studying at a polytechnic or college
(v)   Not studying at a polytechnic or college.

The Probability of Events from Two Trials

THE AVONFORD STAR

A Chance in a Million

I don't know whether Veronica, 14, saw a black cat last Saturday morning, or six magpies, or what but it was certainly her lucky day.

Not only did her one and only ticket win her the top prize of £100 in the raffle at Avonford Summer Fair but later in the afternoon she received another £50 when her programme number came up in the programme draw.

Veronica said, "To be honest, I didn't even know there was a programme draw. I was just about to throw mine away when they announced the winning number. I couldn't believe it – two wins in one day. It must be a chance in a million."

This story describes two pieces of good fortune on the same day. Veronica said the probability was about 1/1 000 000. What was it really?

The two events resulted from two different trials, the raffle draw and the programme draw. Consequently this situation is different from those you met in the previous section. There you were looking at two events from a single trial (like the number coming up when a die is thrown being even or being greater than 4).

The total sales of raffle tickets were 1245 and of programmes 324. The draws were conducted fairly, that is each number had an equal chance of being selected. The table below sets out the two trials and their corresponding events with associated probabilities.

Trial	Events (and estimated probabilities)
Raffle Draw	Winning with a single ticket: 1/1245
	Not winning with a single ticket: 1244/1245
Programme Draw	Winning with a single programme: 1/324
	Not winning with a single programme: 323/324

In situations like this the possible outcomes (events) resulting from the different trials are often shown on a *tree diagram*.

EXAMPLE

Find, in advance of the results of the two draws, the probability that:

(i)   Veronica would win both draws

(ii) Veronica would fail to win either draw.

(iii) Veronica would win one of the two draws.

Solution

These results are shown on the tree diagram, figure 2.5.

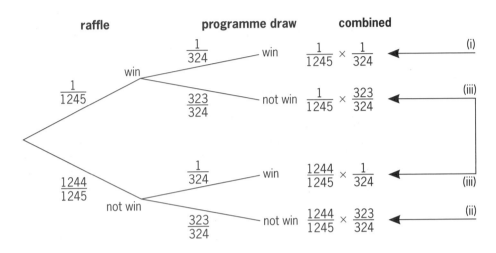

Figure 2.5

(i) The probability that Veronica wins both

$$= \frac{1}{1245} \times \frac{1}{324} = \frac{1}{403\,380}$$

This is not quite Veronica's 'One in a million' but nor is it very far off it.

(ii) The probability that Veronica wins neither

$$= \frac{1244}{1245} \times \frac{323}{324} = \frac{401\,812}{403\,380}$$

This of course is much the most likely outcome.

(iii) The probability that Veronica wins one but not the other is given by

$$\frac{1}{1245} \times \frac{323}{324} + \frac{1244}{1255} \times \frac{1}{324} = \frac{1567}{403\,380}$$

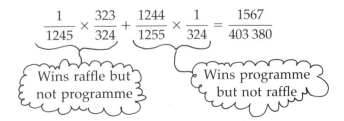

Look again at the structure of the tree diagram (figure 2.5).

There are two trials, the raffle draw and the programme draw. These are considered as *First, Then* trials, and set out *First* on the left and *Then* on the right. Once you understand this, the rest of the layout falls into place, with the different outcomes or events appearing as branches. In this example there are two branches at each stage; sometimes there may be three or more. Notice that for a given trial the component probabilities sum to 1, as before.

$$\frac{1}{403\,380} + \frac{323}{403\,380} + \frac{1244}{403\,380} + \frac{401\,812}{403\,380} = \frac{403\,380}{403\,380} = 1$$

EXAMPLE

Some friends buy a six-pack of potato crisps. Two of the bags are snake flavoured (S), the rest are frog flavoured (F). They decide to allocate the bags by lucky dip. Find the probability that:
(i) the first two bags chosen are the same as each other
(ii) the first two bags chosen are different from each other.

Solution

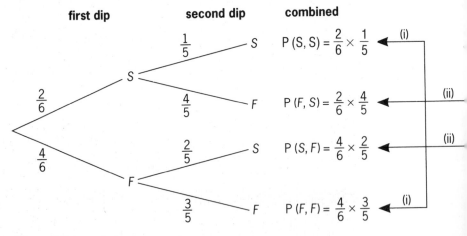

Figure 2.6

(i) P(F, S) means, the probability of drawing a frog bag (F) on the first dip and a snake bag (S) on the second.

The probability that the first two bags chosen are the same as each other = P(S, S) + P(F, F) = 2/6 × 1/5 + 4/6 × 3/5
$$= 1/15 + 6/15$$
$$= 7/15$$

(ii) The probability that the first two bags chosen are different from each other = P(F, S) + P(S, F) = 2/6 × 4/5 + 4/6 × 2/5
$$= 4/15 + 4/15$$
$$= 8/15$$

NOTE

The answer to part (ii) above hinged on the fact that two orderings (S then F, and F then S) are possible for the same combined event (that the two bags selected include one snake and one frog bag). This is a key idea in probability and is the central theme of the next chapter on Permutations and Combinations.

The probabilities changed between the first dip and the second dip. This is because the outcome of the second dip is *dependent* on the outcome of the first one (with fewer bags remaining to choose from).

By contrast, the outcomes of two trials involved in tossing a coin twice are *independent*, and so the probability of getting a head on the second toss remains unchanged at ½, whatever the outcome of the first toss.

Although you may find it helpful to think about combined events in terms of how they would be represented on a tree diagram, you may not always actually draw them in this way. If there are several trials and perhaps more than two possible outcomes from each, drawing a tree diagram can be very time consuming.

Calvin, a keen gambler, writes a regular article in *The Avonford Star* entitled *Calvin's Tips*. This is one of them.

THE AVONFORD STAR

Calvin's tips

I was playing roulette last night. The first time I bet on my lucky number 7 and lucky it was. I then put all my winnings on my other lucky number, 4. That came up too so I collected my winnings and moved off to a poker table.

When I came back later I found my friend Katy still at the roulette table, looking very cross. "My lucky number is 17 and I have tried it 40 times and not once has it come up," she explained. "I am jinxed. I would have expected to win twice by now and that would make me £32 up not £40 down."

I know I was lucky, but was Katy really that unlucky? Did she have good reason to think herself jinxed?

A roulette wheel has 37 numbers marked on it (0, 1, 2, 3, . . ., 36) so the probability that your number comes up on any roll is 1/37.

On one roll,

$$P(\text{win}) = \frac{1}{37} \quad P(\text{not win}) = 1 - \frac{1}{37} = \frac{36}{37}$$

On 40 rolls,

$$P(\text{no wins}) = \frac{36}{37} \times \frac{36}{37} \times \frac{36}{37} \times \ldots \times \frac{36}{37} = \left(\frac{36}{37}\right)^{40}$$

$$= 0.334$$

So no wins in 40 rolls is not particularly unlikely. Katy has no reason to think herself jinxed.

The probability of Calvin's outcome is $\frac{1}{37} \times \frac{1}{37} = 0.00073$. This is a low probability and he is quite right to think himself lucky.

The odds offered by most casinos for a single number are 35–1, meaning that if you bet £1 you win £35 and get your £1 back.

So on one roll the expectation of your winnings is

$$£\left(35 \times \frac{1}{37} - 1 \times \frac{36}{37}\right) = -2.7 \text{ pence, or } 2.7 \text{ pence loss}$$

On 40 rolls Katy could expect to lose 40×2.7 pence $= £1.08$.

In fact she had lost £40 so perhaps she was entitled to think that luck was not on her side, but her statement that she should be £32 in profit was certainly not accurate.

EXAMPLE

THE AVONFORD STAR

Is this justice?

In 1991, local man David Starr was sentenced to 12 years imprisonment for armed robbery solely on the basis of an identification parade. He was one of 12 people in the parade and was picked out by one witness but not by three others.

Many Avonford people who knew David well believe he was incapable of such a crime. We in the Star are now adding our voice to the clamour for a review of his case.

How conclusive, we ask, is this sort of evidence, or to put it another way, how likely is it that a mistake has been made?

Investigate the likelihood that David Starr really did commit the robbery.

Solution
In this situation you need to assess the probability of an innocent individual being picked out by chance alone. Assume that David Starr was innocent and the witnesses were selecting in a purely random way (that is with a probability of 1/12 of selecting each person and a probability of 11/12 of not selecting each person). If each of the witnesses selected just one of the twelve people in the identity parade in this

random manner, how likely is it that David would be picked out by at least one witness?

P(at least one selection) = 1 − P(no selections)

$$= 1 - \frac{11}{12} \times \frac{11}{12} \times \frac{11}{12} \times \frac{11}{12}$$

$$= 1 - 0.706$$

$$= 0.294 \text{ (i.e. roughly 30\%)}$$

In other words, there is about a 30% chance of an innocent person being chosen in this way by at least one of the witnesses. The newspaper article concluded:

This raises an important statistical idea, which you will meet again in Chapter 5, about how we make judgements and decisions.

Judgements are usually made under conditions of uncertainty, and involve us in having to weigh up the plausibility of one explanation against that of another. Statistical judgements are usually made on such a basis. We choose one explanation if we judge the alternative explanation to be sufficiently unlikely, that is if the probability of its being true is sufficiently small. Exactly how small this probability has to be will depend on the individual circumstances and is called the *significance level*.

A criminal on trial is assumed innocent until shown to be guilty beyond reasonable doubt; reasonable doubt must mean a very small probability.

Exercise 2B

1. The probability of a pregnant woman giving birth to a baby girl is about 0.49. Draw a tree diagram showing the possible outcomes if she has two babies (not twins). From the tree diagram, calculate the following probabilities.
   (i)   that the babies are both girls
   (ii) that the babies are the same sex
   (iii) that the second baby is of different sex to the first.

Exercise 2B continued

2. In a certain district of a large city, the probability of a household suffering a break-in in a particular year is 0.07 and the probability of its car being stolen is 0.12. Assuming these two trials are independent of each other, draw a tree diagram showing the possible outcomes for a particular year. Calculate, for a randomly selected household with one car, the following probabilities:
   (i)   that the household is a victim of both crimes during that year.
   (ii) that the household suffers *only one* of these misfortunes during that year.
   (iii) that the household suffers *at least one* of these misfortunes during that year.

3. At an identification parade of 12 people, estimate the probability of the accused person being correctly identified by chance alone. Three witnesses are called to identify the accused person. Assuming they make their choice purely by random selection, draw a tree diagram showing the possible events.

   (a) From the tree diagram, calculate the following probabilities:

   (i)   that all three witnesses select the accused person
   (ii) that none of the witnesses selects the accused person
   (iii) that at least two of the witnesses select the accused person.

   (b) Suppose now that by changing the composition of people in the identification parade, the first two witnesses increase their chances of selecting the accused person to 0.25. Draw a new tree diagram and calculate the following probabilities:

   (i)   that all three witnesses select the accused person
   (ii) that none of the witnesses selects the accused person
   (iii) that at least two of the witnesses select the accused person.

4. Ruth drives her car to work – providing she can get it to start! When she remembers to put the car in the garage the night before, it starts next morning with a probability of 0.95. When she forgets to put the car away, it starts next morning with a probability of 0.75. She remembers to garage her car 90% of the time.

   What is the probability that Ruth drives her car to work on a randomly chosen day?

5. Around 0.8% of men are blue-green colour-blind (the figure is slightly different for women) and roughly 1 in 5 men is left-handed. Assuming these characteristics are inherited independently, calculate with the aid of a tree diagram the probability that a man chosen at random will
   (i) be both colour-blind and left-handed
   (ii) be colour-blind and not left-handed

(iii) be colour-blind or left-handed

(iv) be neither colour-blind nor left-handed.

6. A gambling game consists of tossing a coin three times. You win if all three tosses give the same result (i.e. three heads or three tails) and you lose if any other event shows. Calculate the probability that you will lose a particular game.

7. All the Jacks, Queens and Kings are removed from a pack of cards. Giving the Ace a value of 1, this leaves a pack of 40 cards consisting of four sets of cards numbered 1 to 10. The cards are well shuffled and one is drawn and noted. This card is not returned to the pack and a second card is drawn. Find the probability that
   (i) both cards are even
   (ii) at least one card is odd
   (iii) both cards are of the same suit
   (iv) only one of the cards has a value greater than 7.

8. Three dice are thrown. Find the probability of obtaining
   (i) at least two sixes
   (ii) no sixes
   (iii) different scores on all the dice.

9. Explain the flaw in this argument and rewrite it as a valid statement.
   The probability of throwing a 6 on a fair die = 1/6. Therefore the probability of throwing at least one six in six tosses of the die is $1/6 + 1/6 + 1/6 + 1/6 + 1/6 + 1/6 = 1$ so it is a certainty.

10. In a Donkey Derby event, there are three races. There are 6 donkeys entered for the first race, 4 for the second and 3 for the third. Sheila places a bet on one donkey in each race. She knows nothing about donkeys and chooses each donkey at random. Find the probability that she backs at least one winner.

11. Two dice are thrown, one red and the other green.
    (i) Copy and complete this table showing all the possible outcomes.

		Green die					
		1	2	3	4	5	6
	1						
	2						
Red	3						
die	4						10
	5						11
	6	7	8	9	10	11	12

Exercise 2B continued

    (ii) What is the probability of a score of 4?

    (iii) What is the most likely outcome?

    (iv) Criticise this argument:

        There are 11 possible outcomes, 2, 3, 4 up to 12. Therefore each of them has a probability of 1/11.

12. A gambling game consists of tossing a coin three times. If two or more heads show, you get your money back. If you throw three heads, you receive double your stake money as well. Otherwise you lose. For a single throw, calculate

    (i) the probability of losing your money

    (ii) the probability of doubling your money

    (iii) the probability of just getting your money back

    (iv) the expectation of the value of your winnings, given a £1 stake.

13. The probability of someone catching 'flu' in a particular winter when they have taken the 'flu' vaccine is 0.1. Without the vaccine, the probability of catching 'flu' is 0.4. If 30% of the population have been given the vaccine, what is the probability that a person chosen at random from the population will catch 'flu' over that winter.

14. In a gambling game, players pay an entry fee of £1. They then throw a die until it shows 6. If only one throw is needed, the player receives £3. If two throws are needed, the player receives £2. If three throws are needed, the player gets his or her entry fee back; otherwise the £1 is lost. Find the probability that a player:

    (i) receives £3

    (ii) receives £2

    (iii) gets the £1 entry fee back

    (iv) loses the entry fee.

15. The probability that a girl passes her driving test at any sitting is 2/3. If she fails it, she takes it again, going on until she passes. Find the probability that

    (i) she passes on the third attempt

    (ii) she passes in less than four attempts

    (iii) she takes at least ten attempts to pass.

16. The government of a country is concerned to stabilise the population and passes a law, which is strictly obeyed, that couples should have children until they have a daughter, after which time, no more children are allowed. Find

    (i) the expected number of girls per family

    (ii) the expected number of boys per family

    (iii) the expected sex ratio for the next generation.

Assume that boys and girls are born in equal proportions.

THE AVONFORD STAR

Local man dies of heart attack

Mr. Paul Strangelove, 47, of 23, Dunrobbin Close, collapsed while out shopping late yesterday afternoon. He was rushed to hospital but found to be dead on arrival.

What is the probability that someone chosen at random will die of a heart attack in the next 12 months?

One way of approaching that question would be to say that since there are about 300 000 deaths per year from heart and circulatory diseases (H & CD) among the 57 000 000 population of the UK,

$$\text{probability} = \frac{\text{Number of deaths from H \& CD/year in UK}}{\text{Total population of UK}}$$

$$= \frac{300\,000}{57\,000\,000} = 0.0053$$

However if you think about it, you will probably realise that this is rather a meaningless figure. For a start young people are much less at risk than those in or beyond middle age.

So you might wish to give two answers:

$$P_1 = \frac{\text{Deaths from H \& CD among over-40s}}{\text{Population of over-40s}} \qquad P_2 = \frac{\text{Deaths from H \& CD among under-40s}}{\text{Population of under-40s}}$$

Typically only 1500 of the deaths would be among the under-40s leaving (on the basis of these figures) 298 500 among the over-40s. About 25 000 000 people in the UK are over 40, and 32 000 000 under 40 (40 years and 1 day counts as over 40). This gives

$$P_1 = \frac{\text{Deaths from H \& CD among over-40s}}{\text{Population of over-40s}} = \frac{298\,500}{25\,000\,000}$$

$$= 0.0119$$

and

$$P_2 = \frac{\text{Deaths from H \& CD among under-40s}}{\text{Population of under-40s}} = \frac{1500}{32\,000\,000}$$

$$= 0.00047$$

So someone in the older group is over 200 times more likely to die of a heart attack than someone in the younger group. Lumping them both together as an average figure resulted in a figure that was representative of neither group. (The figures used in this section are approximated from those supplied by the British Heart Foundation who had themselves used a number of official sources.)

But why stop there? You could, if you had the figures, divide the population up into 10 year, 5 year, or even 1 year intervals. That would certainly improve the accuracy; but there are also more factors that you might wish to take into account.

- Is the person overweight?
- Does the person smoke?
- Does the person take regular exercise?
  etc.

The more conditions you build in, the more accurate will be the estimate of the probability.

You can see how the conditions are brought in by looking at P_1:

$$P_1 = \frac{\text{Deaths from H \& CD among over-40s}}{\text{Population of over-40s}} = \frac{298\,500}{25\,000\,000}$$

$$= 0.0119$$

You would write this in symbols as follows:

Event G: Someone selected at random is over 40,
Event H: Someone selected at random dies from H & CD,

The probability of someone dying from H & CD given that he or she is over 40 is given by $P(H|G)$ where

$$P(H|G) = \frac{n(H \cap G)}{n(G)} = \frac{n(H \cap G)/n(\varepsilon)}{n(G)/n(\varepsilon)} = \frac{P(H \cap G).}{P(G)}$$

$P(H|G)$ means the probability of event H occurring given that event G has occurred.

This result may be written in general form for all cases of conditional probability for events A and B

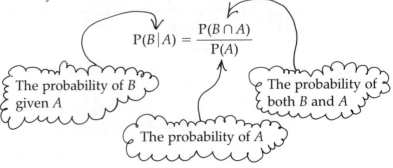

$$P(B|A) = \frac{P(B \cap A)}{P(A)}$$

The probability of B given A

The probability of both B and A

The probability of A

Conditional probability is used when your estimate of the probability of an event is altered by your knowledge of whether some other event has occurred. In this case the estimate of the probability of someone dying from heart and circulatory diseases, $P(H)$, is altered by a knowledge of whether the person is over 40 or not.

Thus conditional probability addresses the question of whether one trial is dependent on another one. If the probability of event B is not affected by the occurrence of event A, we say that B is *independent* of A. If, on the other hand, the probability of event B is affected by the occurrence (or not) of event A, we say that B is *dependent* on A.

If A and B are independent, then $P(B|A) = P(B|A')$ and this is just $P(B)$.

If A and B are dependent, then $P(B|A) \neq P(B|A')$.

As you have already seen, the probability of a combined event is the product of the separate probabilities of each event, provided the question of dependence between the two events is properly dealt with. Specifically:

For dependent events

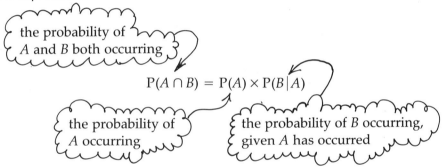

the probability of A and B both occurring

$$P(A \cap B) = P(A) \times P(B|A)$$

the probability of A occurring

the probability of B occurring, given A has occurred

For independent events

When A and B are independent events, then, because $P(B|A) = P(B)$, this can be written as

$$P(A \cap B) = P(A) \times P(B)$$

EXAMPLE

A company is worried about the high turnover of its employees and decides to investigate whether they are more likely to stay if they are given training. On 1st January one year the company employed 256 people (excluding those about to retire). During that year a record was kept of who received training as well as who left the company. The results are summarised in this table:

	Still employed	Left company	
Given training	109	43	152
Not given training	60	44	104
	169	87	256

Find the probability that a randomly selected employee:
(i)   Received training
(ii)   Did not leave the company
(iii) Received training and did not leave the company
(iv)   Did not leave the company, given that the person had received training
(v)   Received training, given that the person had not left the company.

Solution
Using the notation T: The employee received training;
$\qquad\qquad\quad$ S: The employee stayed in the company

(i) $\qquad\qquad P(T) = \dfrac{n(T)}{n(\varepsilon)} = \dfrac{152}{256}$

(ii) $\qquad\qquad P(S) = \dfrac{n(S)}{n(\varepsilon)} = \dfrac{169}{256}$

(iii) $\qquad P(T \cap S) = \dfrac{n(T \cap S)}{n(\varepsilon)} = \dfrac{109}{256}$

(iv) $\qquad P(S \mid T) = \dfrac{P(S \cap T)}{P(T)} = \dfrac{109/256}{152/256} = \dfrac{109}{152}$

(v) $\qquad P(T \mid S) = \dfrac{P(T \cap S)}{P(S)} = \dfrac{109/256}{169/256} = \dfrac{109}{169}$

It is clear that the events T: The employee received training and S: The employee stayed in the company are not independent. Each would seem to be dependent on the other, a conclusion which matches common sense. It is almost certainly true that training increases employees' job satisfaction and so makes them more likely to stay, but it is also probably true that the company is more likely to go to the expense of training the more stable employees who seem less inclined to move on to other jobs.

Exercise 2C

1. In a school of 600 pupils, 360 are girls. There are 320 hockey players, of whom 200 are girls. Among the hockey players there are 28 goalkeepers, 19 of them girls. Find the probabilities that:
   (i)   a pupil chosen at random is a girl
   (ii)   a girl chosen at random plays hockey
   (iii)   a hockey player chosen at random is a girl
   (iv)   a pupil chosen at random is a goalkeeper
   (v)   a goalkeeper chosen at random is a boy
   (vi)   a boy hockey player chosen at random is a goalkeeper
   (vii)   a hockey player chosen at random is a boy goalkeeper
   (viii) two pupils chosen at random are both goalkeepers

(ix) two pupils chosen at random are a boy goalkeeper and a girl goalkeeper

(x) two pupils chosen at random are one boy and one girl.

2. The table below gives the numbers of offenders in England and Wales sentenced for indictable offences, by type of offence and type of sentence, in 1989 (numbers in thousands)

	Discharge	Probation	Fine	Jail	Other	Total
Violence	9.5	3.9	20.6	9.0	12.7	55.7
Sexual offences	0.8	0.9	2.4	2.4	0.8	7.3
Theft	31.4	26.2	62.6	29.9	32.1	182.2
Motoring	0.5	0.3	8.7	1.0	0.8	11.3
Other	12.0	9.4	41.3	11.9	7.9	82.5
Total	54.2	40.7	135.6	54.2	54.3	339.0

Source: adapted from Table 12.19, *Social Trends* 22, 1992, CSO.

(a) Find the probability that a randomly selected person indicted for an offence will be

(i)   discharged   (ii)   put on probation
(iii) fined   (iv)   sent to jail
(v)   sent to jail for a motoring offence
(vi) sent to jail given that the person has committed a motoring offence
(vii) guilty of a motoring offence given that the person is sent to jail.

(b) Criticise this statement:
On these figures nearly 2% of the country's prison population are there for motoring offences.

3. 100 cars are entered for a road-worthiness test which is in two parts, Mechanical and Electrical. A car passes only if it passes in both parts. Half the cars fail the electrical test and 62 pass the mechanical. 15 pass the electrical but fail the mechanical test. Find the probability that a car chosen at random
(i)   passes overall (ii) fails on one test only
(iii) given that it has failed, failed the mechanical test, only.

4. Two dice are thrown. What is the probability that the total is:
(i) 7; (ii) a prime number; (iii) 7, given that it is a prime number?

5. A cage holds two litters of rats. One litter comprises 3 females and 4 males, and the other comprises 2 females and 6 males. A random selection of 2 rats is made. Find, as fractions, the probabilities that the 2 rats are
(i)   from the same litter,

Exercise 2C continued

    (ii) of the same sex,

    (iii) from the same litter and of the same sex,

    (iv) from the same litter given that they are of the same sex.

[MEI]

6. In a school of 400 pupils, 250 play a musical instrument and 100 sing in the choir.

   The probability that a pupil chosen at random neither plays a musical instrument, nor sings in the choir is 1/5.

   Find

   (i)   how many pupils both sing in the choir and play a musical instrument

   (ii) the probability that a pupil chosen at random sings in the choir but does not play an instrument

   (iii) the probability that a member of the choir chosen at random does not play an instrument

   (iv) the probability that someone who does not play an instrument, chosen at random, is in the choir.

7. A bag P contains 3 red balls. A second bag Q contains 2 red balls and 3 black balls.
   (a) A bag is chosen at random and one ball is withdrawn. Find the probability that this ball is red.
   This ball remains outside the bag.
   (b) A bag is again chosen at random (it is not known whether this is the same bag as before or not) and one ball is withdrawn. Find the joint probability that both this ball and the one previously withdrawn are red.
   (c) If they are both red, what is the probability that bag P was used on both occasions?

[O & C]

8. In a child's game there should be 7 triangles, 3 of which are blue and 4 of which are red, and 11 squares, 5 of which are blue and 6 of which are red. However, two pieces are lost. Assuming the pieces are lost at random, find the probability that they are
   (i)   the same shape,
   (ii) the same colour,
   (iii) the same shape and the same colour,
   (iv) the same shape given that they are the same colour.

[MEI]

9. A and B are two events with probabilities given by
   $P(A) = 0.4$, $P(B) = 0.7$ and $P(A \cap B) = 0.35$.
   (i) Find $P(A|B)$ and $P(B|A)$
   (ii) Show that the events A and B are not independent

10. Quark hunting is a dangerous occupation. On a quark hunt, there is a probability of ¼ that the hunter is killed. The quark is twice as likely to be killed as the hunter. There is a probability of ⅓ that both survive.

(i) Copy and complete this table of probabilities

	Hunter dies	Hunter lives
Quark dies		½
Quark lives	⅓	½
	¼	1

Find the probability that

(ii) both the hunter and the quark die

(ii) the hunter lives and the quark dies

(iii) the hunter lives, given that the quark dies.

KEY POINTS

- The probability of an event A, $P(A) = n(A)/n(\varepsilon)$, where $n(A)$ is the number of ways that A can occur and $n(\varepsilon)$ is the total number of ways that all possible events can occur, all of which are equally likely.

- For any two events, A and B, of the same trial,

$$P(A \cup B) = P(A) + P(B) - P(A \cap B).$$

Where the events do not overlap (i.e. where the events are mutually exclusive), the rule still holds but, since $P(A \cap B)$ is now equal to zero, the equation simplifies to:

$$P(A \cup B) = P(A) + P(B)$$

- Where a trial produces two or more mutually exclusive events, the probabilities of the separate events sum to 1.

continued

Statistics 1

2

- $P(A) + P(A') = 1$

- $P(B|A)$ mean the probability of event B occurring, given that event A has already occurred.

$$P(B|A) = \frac{P(A \cap B)}{P(A)}$$

The probability that event A and then event B occur, in that order, is $P(A) \times P(B|A)$

- If event B is independent of event A, $P(B|A) = P(B|A') = P(B)$.

3

Selections

As I was going to St. Ives,
I met a man with seven wives.
Each wife had seven sacks;
Each sack had seven cats;
Each cat had seven kittens.
Kittens, cats, sacks and wives,
How many were going to St. Ives?

THE AVONFORD STAR

Child Prodigy

Little Gary Forest looks like any other toddler but all the evidence points to him being a budding genius.

Recently Gary's mother, attractive brunette Michelle, gave him five bricks to play with, each showing one of the numbers **1**, **2**, **3**, **4** and **5**. Without hesitation Gary sat down and placed them in the correct order.

Little Gary Forest looks like being a budding genius!

What do you think?

Look at the diagram.

| 1 | 2 | 3 | 4 | 5 |

If Gary had actually chosen them at random:

> the probability of first selecting **1** is 1/5
> the probability of next selecting **2** is 1/4
> the probability of next selecting **3** is 1/3
> the probability of next selecting **4** is 1/2
> then only **5** remains so the probability of selecting it is 1

So the probability of getting the correct numerical sequence at random is

$$\frac{1}{5} \times \frac{1}{4} \times \frac{1}{3} \times \frac{1}{2} \times 1 = \frac{1}{120}$$

Do you agree with *The Avonford Star* that Gary is a child prodigy, or do you think it was just by chance that he put the bricks down in the right order?

What more information would you want to be convinced that he is a budding genius?

Factorials

In the last example you encountered the term $5 \times 4 \times 3 \times 2 \times 1$ on the bottom line of the calculation. It is called *5 factorial* and it is written 5!. You will often meet expressions of this form.

In general

$$n! = n \times (n-1) \times (n-2) \times \ldots \times 3 \times 2 \times 1$$

where n is a positive integer.

EXAMPLE | Calculate 7!

Solution $7! = 7 \times 6 \times 5 \times 4 \times 3 \times 2 \times 1 = 5040$

Some typical relationships between factorial numbers are illustrated below:

$10! = 10 \times 9!$ or in general $n! = n \times (n-1)!$
$10! = 10 \times 9 \times 8 \times 7!$ or in general $n! = n \times (n-1) \times (n-2) \times (n-3)!$

These are useful when simplifying expressions involving factorials.

EXAMPLE	Calculate $\dfrac{5!}{3!}$

Solution: $\dfrac{5!}{3!} = \dfrac{5 \times 4 \times 3!}{3!} = 5 \times 4 = 20$

EXAMPLE	Calculate $\dfrac{7! \times 5!}{3! \times 4!}$

Solution:

$$\dfrac{7! \times 5!}{3! \times 4!} = \dfrac{7 \times 6 \times 5 \times 4 \times 3! \times 5 \times 4!}{3! \times 4!} = 7 \times 6 \times 5 \times 4 \times 5 = 4200$$

EXAMPLE	Write in terms of factorials only $37 \times 36 \times 35$.

Solution:

$$37 \times 36 \times 35 = \dfrac{37 \times 36 \times 35 \times 34!}{34!} = \dfrac{37!}{34!}$$

Exercise 3A

1. Calculate (i) $8!$ (ii) $\dfrac{8!}{6!}$ (iii) $\dfrac{5! \times 6!}{7! \times 4!}$

2. Calculate (i) $\dfrac{(n-1)!}{n!}$ (ii) $\dfrac{(n-1)!}{(n-2)!}$

3. Calculate (i) $\dfrac{(n+3)!}{(n+1)!}$ (ii) $\dfrac{n!}{(n-2)!}$

4. Write in factorial notation:

(i) $\dfrac{8 \times 7 \times 6}{5 \times 4 \times 3}$ (ii) $\dfrac{15 \times 16}{4 \times 3 \times 2}$ (iii) $\dfrac{(n+1)\,n(n-1)}{4 \times 3 \times 2}$

5. Factorise:

(i) $7! + 8!$ (ii) $n! + (n+1)!$

6. How many different four letter words can be formed from the letters A, B, C, D if letters cannot be repeated? (The words do not need to mean anything)

Exercise 3A continued

7. How many different four letter words can be formed from the letters A, B, C, D, E, F if letters cannot be repeated?

8. How many different ways can eight books be arranged in a row on a shelf?

9. In a greyhound race there are six runners. How many different ways are there for the six dogs to finish?

10. Simplify $\dfrac{(n + m + 1)!}{(n + m - 1)!\,(n + m + 1)}$

11. Solve the inequality $n! > 10^m$ for each of the cases $m = 3, 4, 5$.

Investigation

In how many ways can you write 42 in terms of factorials only?

Several Events

There are three roads from Adamsville to Brantwood and two roads from Brantwood to Chandford. How many road routes are there from Adamsville to Chandford passing through Brantwood on the way?

Look at figure 3.1.

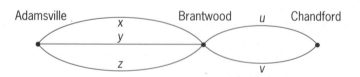

Figure 3.1

The solution to our problem is $3 \times 2 = 6$ because the possible routes are

$$x—u,\ y—u,\ z—u,\ x—v,\ y—v,\ z—v.$$

In general, if there are m routes connecting town A with town B and n routes connecting town B with town C then there are $m \times n$ ways to travel from town A to town C via town B.

There are also four roads from Chandford to Digby. How many different routes are there from Adamsville to Digby travelling via Brantwood and Chandford?

Look at figure 3.2.

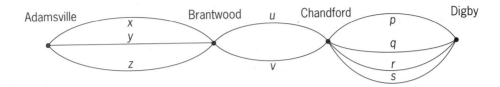

Figure 3.2

Extending the last result we can say that the number of required routes $= 3 \times 2 \times 4 = 24$. They can be listed systematically as follows:

$$x-u-p \quad y-u-p \quad z-u-p \quad x-v-p \quad y-v-p \quad z-v-p$$
$$x-u-q \quad \cdots \cdots \quad \cdots \cdots \quad \cdots \cdots \quad \cdots \cdots \quad \cdots \cdots$$
$$x-u-r \quad \cdots \cdots \quad \cdots \cdots \quad \cdots \cdots \quad \cdots \cdots \quad \cdots \cdots$$
$$x-u-s \quad \cdots \cdots \quad \cdots \cdots \quad \cdots \cdots \quad \cdots \cdots \quad z-v-s$$

In general, if there are a ways for trial A to result, b ways for trial B to result and c ways for trial C to result then there are $a \times b \times c$ different possible outcomes of the three trials.

EXAMPLE

In a 60 metre hurdles race there are 5 runners, one from each of the nations Austria, Belgium, Canada, Denmark and England. How many different finishing orders are there?

Solution:
Let us call the runners A, B, C, D and E.
The winner can be chosen in 5 ways (either A, B, C, D or E).
The runner-up can be chosen in 4 ways (any of A, B, C, D or E omitting the winner).
The third place can be chosen in 3 ways (any of A, B, C, D, or E bar the two runners already home).
The fourth position can be chosen in 2 ways.
The final person can be chosen in 1 way only.
Therefore the total different number of finishing orders

$$= 5 \times 4 \times 3 \times 2 \times 1 = 5! = 120$$

We may generalise this result by saying that the number of ways of arranging n unlike objects in a line is $n!$

EXAMPLE

A £1 coin is tossed 4 times and the result recorded in the form *HTTH*. How many different four letter sequences give the possible results?

Solution:
Each toss of the coin has 2 possible outcomes, heads or tails. Therefore, the total number of possible sequences is $2 \times 2 \times 2 \times 2 = 16$.

EXAMPLE

How many different vehicle registration plates can be formed if each is to contains the letters A, B, C, D and the three digits 2, 6 and 9? The number plate must start with a letter, followed by three digits, and end with three letters.

Solution:
The first letter can be chosen in 4 ways.
The three digits can be arranged in 3! ways = 6 ways
The final three letters can be arranged in 3! ways = 6 ways
Therefore total number of ways = $4 \times 6 \times 6 = 144$.

Investigation:

Number plates in the United Kingdom currently consist of a 7 character registration made up of a letter followed by three digits and then three letters at the end.
(i) How many different number plates can be issued according to this set of rules?
(ii) Is it better to avoid using certain letters of the alphabet and if so how does this affect your answer to the first part?
(You will find it simplest in this investigation to avoid old number plates issued before these rules came in, and special plates like those issued by the Army.)

Exercise 3B

1. At a school there are 13 applicants for a Biology post and 14 applicants for a French post. How many different ways are there of filling the two vacancies?

2. My brother Mark fancies himself as an impresario and decides to put together a rock band from amongst his year at school. He wants a lead singer, a guitarist, a keyboard player and a drummer. He invites applications and gets 7 singers, 5 guitarists, 4 keyboard players and 2 drummers. Assuming each person applies only once, in how many ways can Mark put the group together?

3. From a sixth form of 45 boys and 37 girls a head boy and head girl have to be chosen. How many possible selections are there?

4. How many different three letter words can be formed from the letters A, B, C, D, E if (i) letters may not be repeated; (ii) letters may be repeated?

5. A football match can result in one of three possible outcomes: a home win, an away win or a draw. How many possible outcomes are there to six football matches?

6. Dawn opens a bank account and receives a cash card. She is allowed to make up her own 4 digit code number that will allow her to use the card.

   If she restricts her choice to the numbers 2, 3, 4 and 5 in how many ways can she make up the number, if she uses each number only once?

   If she uses the numbers 1 to 9, and again uses each number only once, in how many ways can she make up the number?

   Using the numbers 1 to 9, in how many ways can she make up the number if she allows repeats?

   Dawn decides that she will use the digits 0 to 9 with repeats allowed, but the bank will not allow a number that starts with 0; in how many ways can the number be made up?

7. In the game of Ludo a six has to be thrown with an ordinary die in order for a counter to commence its journey round the board. What is the probability that (i) a counter starts on the fourth turn; (ii) a counter requires more than four turns to start?

8. Find how many teams of two pairs (1st and 2nd), each pair consisting of a man and a woman, can be chosen to represent a club at a badminton match, if these pairs are chosen from five men and three women.

9. The last two questions of an examination paper are multiple choice. There is one correct answer and five wrong answers to each question. In how many ways is it possible to select the wrong answer to both questions?

EXAMPLE   Suppose you have four books, consisting of one copy each of *David Copperfield* and *Nicholas Nickleby* and two copies of *Little Dorrit*. In how many different ways can you arrange the titles on a bookshelf?

Solution:
Label the titles: *D* (*David Copperfield*), *N* (*Nicholas Nickleby*) and *L* (*Little Dorrit*). The possible arrangements are:

LLDN, LLND, LDLN, LNLD, LDNL, LNDL, DLLN, NLLD, DLNL, NLDL, DNLL, NDLL.

Thus there are 12 possible arrangements.

Now if you had had four different books there would have been 4! = 24 different ways to arrange them in a line.

Can we connect the two answers? It seems that having two copies of one of the books has halved the total number of arrangements.

EXAMPLE There are four books consisting of one copy of *David Copperfield* and three copies of *Little Dorrit*. In how many different ways can you arrange them on a bookshelf?

Solution:
Using the same notation as in the previous example, the possible arrangements are:

DLLL, LDLL, LLDL, LLLD.

Thus there are only 4 arrangements in this case.

Can you see a connection between the number of arrangements in this case and in the case when the four books are all different?

When you had 2 copies of *Little Dorrit*, the number of arrangements

$$= \frac{4!}{2!} = 12$$

When you had 3 copies of *Little Dorrit*, the number of arrangements

$$= \frac{4!}{3!} = 4$$

We may generalise this result by saying the number of different ways of arranging n objects of which a are alike is

$$\frac{n!}{a!}$$

EXAMPLE There are four books consisting of two copies of each of *David Copperfield* and *Little Dorrit*. In how many different ways can we arrange the titles on a bookshelf?

Solution:

Using the same notation as before the possible arrangements are:

DDLL, DLDL, DLLD, LDDL, LDLD, LLDD.

Thus there are 6 arrangements.

How is this connected to the number of ways of arranging 4 different books?

A little thought shows that $\dfrac{4!}{2! \times 2!} = 6$

This result can be generalised as follows:

the number of different ways of arranging n objects of which a are of one type, b are of a second type, c are of a third type, etc. is

$$\frac{n!}{a!\ b!\ c!\ \ldots}$$

EXAMPLE   In how many ways can the letters of the word RECTANGLE be arranged?

Solution:
RECTANGLE has 9 letters with 2 Es

The number of arrangements $= \dfrac{9!}{2!} = 181\,440$

EXAMPLE   In how many ways can the letters of the word PROBABILITY be arranged?

Solution:
PROBABILITY has 11 letters including 2 Bs and 2 Is.

The number of arrangements $= \dfrac{11!}{2!\,2!} = 9\,979\,200$

EXAMPLE   In how many ways can the letters of the word PROBABILITY be arranged if the first and last letters must be R and B respectively?

Solution:
Since the first and last letters are stipulated we have only to arrange the middle 9 letters. These letters are POABILITY.

We have a 9 letter word with two Is, so there are

$$\frac{9!}{2!} = 181\,440 \text{ ways.}$$

Exercise 3C

1. How many distinct arrangements are there of the letters of the words
   (i) WARREN; (ii) PRINCIPLE?

2. In how many different ways can the letters of the word STATISTICS be arranged?

3. (i) How many distinct arrangements are there of the letters of the word COMMITTEE?
   (ii) How many arrangements will there be if the first and last letters are C and E respectively?

4. How many distinct arrangements are there of the letters of the word POSSESSES?

5. Three letters are selected at random from the word GEOGRAPHY. Find the probability that the selection (i) does not contain the letter G; (ii) contains both the letter Gs.

6. If the letters of the word CHEMISTRY are arranged at random, find the probability that the arrangement ends in the letter T.

7. How many different words of four letters can be formed from the word REGAL?

8. How many numbers between 10 and 500 can be made from the digits 1, 2, 5, 5 if (i) each digit may be used only once; (ii) each digit may be used more than once?

9. Two sets of books contain six different Mathematics books and four different French books respectively. In how many ways can the books be arranged on a shelf if the Mathematics and French books are not mixed up?

10. How many different words of three letters can be formed from the word BROWN if each word must contain a vowel?

Permutations

EXAMPLE Ten athletes compete in the 1500 m final. The winner receives a gold medal, the runner-up a silver medal and the third person a bronze medal. In how many different ways can the medals be awarded?

Solution:
the winner could be selected in any of 10 ways
the second could be selected in any of 9 ways
the third could be chosen in any of 8 ways

Thus total number of ways $= 10 \times 9 \times 8 = 720$.

In this example attention is given to the order in which the competitors finish. The solution required a *permutation* of three objects from ten.

In general the number of permutations, ^nP_r, of r objects from n is given by

$$^nP_r = n \times (n-1) \times (n-2) \times \ldots \times (n-r+1)$$

This can be written more compactly as

$$^nP_r = \frac{n!}{(n-r)!}$$

In the example of the athletes in the 1500 m final this was

$$^{10}P_3 = \frac{10!}{7!} = 720 \text{ ways.}$$

EXAMPLE Four letters are picked from the word CONSIDER. In how many ways can this be done?

Solution:
This is an arrangement of 4 out of 8 different objects.

$$^8P_4 = \frac{8!}{(8-4)!} = 8 \times 7 \times 6 \times 5 = 1680 \text{ ways}$$

Combinations

Very often we are not concerned with the order in which the items are chosen but only with which items are picked. Suppose that the 1500 m race mentioned in the last section was not the final but only a heat of ten runners, where the first three athletes home go through into the final. How many different selections are possible?

The relative positions of the first three finishers are not important. We are searching for the number of *combinations* of three items from 10. Now each combination of three members can be ordered in 3! ways so

3! × number of combinations = number of permutations.

Therefore,

$$\text{number of combinations} = \frac{\text{number of permutations}}{3!}$$

$$= \frac{^{10}P_3}{3!} = \frac{10!}{7!\,3!} = 120 \text{ ways}$$

In general the number of selections or combinations of r objects from n is denoted by ^nC_r or $\begin{pmatrix} n \\ r \end{pmatrix}$

and this is given by $^nC_r = \dfrac{n!}{(n-r)!\,r!}$

EXAMPLE

A School Governors' committee of 5 people is to be chosen from 8 applicants. How many different selections are possible?

Solution:

$$\text{Number of selections} = {}^8C_5 = \frac{8!}{3! \times 5!} = \frac{8 \times 7 \times 6}{3 \times 2 \times 1} = 56.$$

EXAMPLE

A cricket team consisting of 6 batsmen, 4 bowlers and 1 wicket-keeper is to be selected from a group of 18 cricketers comprising 9 batsmen, 7 bowlers and 2 wicket-keepers. How many different teams can be selected?

Solution:
The batsmen can be selected in $^9C_6$ ways.

The bowlers can be selected in $^7C_4$ ways.

The wicket-keepers can be selected in $^2C_1$ ways.

Therefore total number of teams $= {}^9C_6 \times {}^7C_4 \times {}^2C_1$

$$= \frac{9!}{3!\,6!} \times \frac{7!}{3!\,4!} \times \frac{2!}{1!\,1!}$$

$$= \frac{9 \times 8 \times 7}{3 \times 2 \times 1} \times \frac{7 \times 6 \times 5}{3 \times 2 \times 1} \times 2$$

$$= 5880 \text{ ways.}$$

EXAMPLE

In how many ways can a committee of 4 people be selected from 4 applicants?

Solution:
Common sense tells us that there is only 1 way to make the committee. That is by appointing all applicants. However, if we work from the formula

$$^{4}C_{4} = \frac{4!}{0!\,4!} = \frac{1}{0!}$$

So we use the convention that 0! is taken to be 1 in order to achieve the correct answer to this and other problems involving factorials.

EXAMPLE

A group of 5 computer programmers is to be chosen to form the night shift from a set of 14 programmers. In how many ways can the programmers be chosen if the 5 chosen must include the shift-leader who is one of the 14?

Solution:
Since one place must go to the shift-leader, we have the freedom to select 4 programmers from the remaining 13.

$$\text{Number of selections} = {}^{13}C_{4} = \frac{13!}{9! \times 4!} = \frac{13 \times 12 \times 11 \times 10}{4 \times 3 \times 2 \times 1} = 715$$

Combinations are an important topic and will be met again in the chapter on the binomial distribution. In particular expressions like $^{5}C_{0}$, $^{5}C_{1}$, $^{5}C_{2}$, $^{5}C_{3}$, $^{5}C_{4}$, $^{5}C_{5}$ are found to occur in the expansion of $(1 + x)^{5}$:

$$(1 + x)^{5} = {}^{5}C_{0} + {}^{5}C_{1}x + {}^{5}C_{2}x^{2} + {}^{5}C_{3}x^{3} + {}^{5}C_{4}x^{4} + {}^{5}C_{5}x^{5}$$

THE AVONFORD STAR

Calvin's Tips

When you are playing poker, you should always take good care over where you are sitting. Don't allow any spectators behind you; the other players will watch them to see if they can tell anything about your hand from their faces. I always try to sit opposite the strongest player at the table.

As a mathematical exercise you might like to work out how many ways four people can arrange themselves at a round table. Assume that all the chairs are identical and that there are no peculiarities of any seating position.

One of those will give you the best chance of winning.

Begin by placing Calvin in a particular position. This leaves 3 people to arrange around him, see figure 3.3

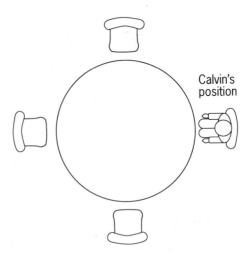

Calvin's position

Figure 3.3

Imagine that Calvin sits down first and that the other seats are then filled in turn going round from his left.

First Calvin sits anywhere. Since all places are the same this counts as 1 way.   1

There are 3 ways of filling the place on Calvin's left.   3

That leaves 2 people for the seat opposite Calvin.   2

There is only one person left to fill the seat on Calvin's right.   1

So the total number of ways is $1 \times 3 \times 2 \times 1 = 6$.

It is easy to see that this result can be generalised to n people (or objects). The number of ways of arranging n different objects in a circle is $(n-1)!$

EXAMPLE   Eight people, 4 boys and 4 girls, are to sit at a round table.

(i) In how many different ways can this be done?

(ii) If they sit down at random, what is the probability that boys and girls will be seated alternately?

Solution:

(i) Number of ways to seat 8 people at a round table $= (8-1)!$
$$= 5040.$$

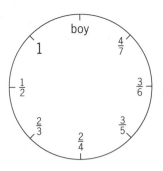

Figure 3.4

(ii) Let us say that a particular boy is seated at the 12 o'clock position as in figure 3.4.

Probability that a girl sits to his left = 4/7

Probability that a boy sits to her left = 3/6

Probability that a girl sits to his left = 3/5

Probability that a boy sits to her left = 2/4

Probability that a girl sits to his left = 2/3

Probability that a boy sits to her left = 1/2

Probability that a girl sits in last place = 1

$$\text{Therefore probability} = \frac{4}{7} \times \frac{3}{6} \times \frac{3}{5} \times \frac{2}{4} \times \frac{2}{3} \times \frac{1}{2} \times 1$$

$$= \frac{1}{35}$$

Exercise 3D

1. In how many ways can 6 boys seat themselves at a round table?

2. (i) How many 5 digit odd numbers can be formed with the digits 1, 2, 3, 4, 5 if repetitions of digits are not allowed?
   (ii) How many 5 digit odd numbers can be formed if repetitions of the digits are allowed?

3. From a group of 6 boys and 5 girls, 2 pupils are chosen at random. Find the probability that there will be one boy and one girl.

4. How many even 5 digit numbers can be formed from the digits 0 to 9 if no digit can occur more than once?

Exercise 3D continued

5. How many different selections of 4 cards can be made from a pack of 52? How many of them will include the Ace of Spades?

6. Anne has five coins and Brian has three. They while away their time waiting for their maths teacher by tossing them. What is the probability that they will both toss the same number of tails?

7. Mrs Smale is having trouble getting the girls' hockey team to turn up on time for training. She decides that as a reward she will take the first 6 who turn up tomorrow to see an international match. If there are 20 girls in the squad, in how many ways can the 6 be chosen?

   If the squad is made up of 2 goalkeepers, 10 defenders and 8 attackers what is the probability that the lucky 6 will not include any attackers?

8. Emma takes her three young brothers for a ride on the Severn Valley Railway. They get into an empty compartment with 3 forward-facing and 3 rearward-facing seats.
   (i) In how many different ways can they be seated?
   (ii) In how many ways will two particular boys be sitting opposite each other?

9. After his first day studying for A levels Sunil finds he has 5 Maths books and 5 Geography books.

   In how many ways can the books be arranged on his bookshelf if he just puts them there in any order?

   If he decides to put the Maths books together, and the Geography books together, in how many ways can they now be arranged?

   Sunil's young sister decides to 'help' by rearranging his books. In how many ways can she do this if she decides to alternate the books of each subject?

10. A class contains 24 children, 14 boys and 10 girls. Four complimentary tickets are given at random to children in the class to attend a soccer match. What is the probability that
    (i) all 4 tickets go to girls,
    (ii) 2 boys and 2 girls receive tickets?

11. My parents are celebrating their golden wedding anniversary and we are going to have a big family dinner for them at home. The dining-table will seat only 12, but there are 17 members of the family who live near enough to be invited. Ignoring any restrictions on choice, in how many ways can I invite 12 members of the family?

    Of course I must invite my parents, and my brother and I must be there; in how many ways can I now invite the 12?

    If I invite Aunty Joyce, I must invite her husband Reg; in how many ways can I now invite everyone?

Exercise 3D continued

Having finally chosen the 12 who will be present at the dinner, they now have to be seated at the large round dining table. In how many ways can they all sit down to dinner?

After much debate we have decided to invite Reg and Joyce. They are such a nuisance that it would be best to sit them down together. In how many ways can the party now be seated?

Investigation

Investigate the mathematics behind the football pools.

You have to choose 8 matches, called a line, for each of which you score points:

Score draw	3 points
No-score draw	2 points
Away win	1.5 points
Home win	1 point

The biggest payout goes to the person with the most points.

Some people choose more than 8 matches and then want all possible selections of 8 considered from among those they have picked, each one a separate line. Others want certain matches (bankers) included in all their lines. (Note that when the pools companies talk about 'perms' they don't actually mean permutations but combinations.)

KEY POINTS

- $n! = n \times (n-1) \times (n-2) \times (n-3) \times \ldots \times 3 \times 2 \times 1$
- The number of ways of arranging n unlike objects in a line is $n!$
- The number of ways of arranging n objects of which a are of one type, b are of a second type, c are of a third type, etc is

$$\frac{n!}{a!\; b!\; c! \;\ldots}$$

- The number of permutations of r objects from n is

$$^nP_r = \frac{n!}{(n-r)!}$$

- The number of combinations of r objects from n is

$$^nC_r = \frac{n!}{(n-r)!\, r!}$$

- The number of ways of arranging n unlike objects in a circle is $(n-1)!$

Exercise 3E

1. Calculate (i) 10!   (ii) $\dfrac{6! \; 7!}{3! \; 4!}$

2. Calculate (i) $\dfrac{(k+5)!}{(k+3)!}$ (ii) $\dfrac{(r+s+2)!}{(r+s)! \, (r+s+2)}$

3. In how many different ways can 5 people sit at a round table?

4. In how many ways can seven different coloured beads be arranged on a ring?

5. How many distinct arrangements are there of the letters of the words (i) CENTURY; (ii) MILLIMETRE?

6. How many odd numbers can be formed from the digits 2, 3, 4, 5, 6 and 7 if repetitions are not allowed?

7. In a game of bridge the pack of 52 cards is shared equally between four players. What is the probability that a particular player has no picture cards (Jacks, Queens, Kings)?

8. How many arrangements can be made of four letters chosen from PLEAT if the first letter is a consonant and each arrangement contains four different letters?

9. Find the number of integers between 1000 and 4000 which can be formed using the digits 1, 2, 3, 4
   (i) if each digit may be used only once,
   (ii) if each digit may be used more than once.

10. How many different hands of four cards can be dealt from a pack of 52 playing cards if at least one of the cards is a picture (Jack, Queen or King)?

11. A team of 4 is chosen at random from 7 girls and 6 boys; in how many ways can the team be chosen if:
    (i) there are no restrictions,
    (ii) there must be more girls than boys?

12. In the Premier league there are 7 matches. If each match can result in a home win, an away win or a draw find the total number of possible results for these fixtures.

13. From a group of 7 men and 5 women, 5 people are selected at random. Find the probability that there are more women than men.

14. A touring party of hockey players is made up of 5 players from each of England, Ireland and Scotland, and 3 from Wales.
    (i) How many different selections of 11 players can be made for a team?
    (ii) In one match, it is decided to have 3 players from each of England, Ireland and Scotland and 2 from Wales. How many different team selections can now be made?
    (iii) In the next match, it is decided that the team should have the same composition of the various nationalities but that the captain, who is from Scotland, must play, as must the better goal-keeper, who is from Wales. The reserve goal-keeper, from Scotland, cannot play. How many team selections can now be made?
    (iv) At the end of the tour three medals, gold, silver and bronze, are to be awarded to the best, second best and third best players respectively. In how many different ways can the three medals be given?

15. I have a box of chocolates with 10 different chocolates left in it. Of these, there are 6 which I particularly like. However, I intend to offer my three friends one chocolate each before I eat the rest. How many different selections of chocolates can I be left with after my friends have chosen?

    Show that 36 of these selections leave me with exactly 5 chocolates which I particularly like.

    How many selections leave me with (i) all 6; (ii) exactly 4; and (iii) exactly 3 of the chocolates which I particularly like?

    Assuming that my friends choose at random, what is the most likely outcome and what is the probability of that outcome?

    [MEI]

16. A committee of four is to be selected from 10 candidates, 6 men (including Mr Brown) and 4 women (including Mrs Smith). In how many distinct ways can the committee be chosen?

    Assuming that each candidate is equally likely to be selected, determine the probabilities that the chosen committee contains
    (i) no women
    (ii) two men and two women
    (iii) Mr Brown, another man and any two women
    (iv) Mrs Smith and any three others, excluding Mr Brown.

    [MEI]

17. In a certain code, each letter of an alphabet is represented by a sequence of symbols. Each symbol is a dot or a dash. Find how many different letters can be represented by a sequence of
    (i) exactly four symbols,
    (ii) not more than three symbols.

    [C]

Exercise 3E continued

18. I recently had a pair of doors made by a joiner. Each door has 15 panes of glass arranged as shown below. The panes marked * have identical patterns on them, and all other panes are plain.

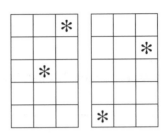

(i)   In how many different ways could the joiner position the 2 patterned panes in the left hand door?

(ii)   In how many different ways could the joiner position the 4 patterned panes, 2 in each door?

Suppose I specified that the 2 patterned panes in the left hand door should not appear in the same row or column.

(iii) After the joiner has positioned the first patterned pane in the left hand door, how many places are there where he could position the second?

(iv)   Deduce that there are 60 different ways in which he could build the left hand door to my specification. Explain your reasoning carefully.

Now consider the pair of doors as an array of 5 rows and 6 columns of panes.

(v)   In how many different ways could the joiner position the 4 patterned panes, 2 in each door, if I specified that no two panes should appear in the same row or column? (Note that the diagram shows one arrangement which meets this specification.)

[ME

4 The Binomial Distribution

To be or not to be, That is the question.

Shakespeare (Hamlet)

Samantha's great invention

Mother of three Samantha Weeks has done more than her bit to protect the environment. She has invented the first cheap energy-saving light-bulb.

Now Samantha is out to prove that she is not only a clever scientist but a smart business woman as well. For Samantha is setting up her own factory to make and sell her SUPERSAVER bulbs.

Samantha admits there are still some technical problems . . .

Samantha Weeks hopes to make a big success of her light industry

Samantha's production process is indeed not very good and there is a probability of 0.1 that any bulb will be substandard, and so not last as long as it should.

She decides to sell her bulbs in packs of three and believes that if one bulb in a pack is substandard, the customers will not complain but that if two or more are substandard they will do so. She also believes that complaints should be kept down to no more than 2.5% of customers. Does she meet her target?

Imagine a pack of Samantha's bulbs. There are eight different ways that good (G) and substandard (S) bulbs can be arranged in Samantha's packs, each with its associated probability.

Arrangement			Probability	Good	Substandard
G	G	G	$0.9 \times 0.9 \times 0.9 = 0.729$	3	0
G	G	S	$0.9 \times 0.9 \times 0.1 = 0.081$	2	1
G	S	G	$0.9 \times 0.1 \times 0.9 = 0.081$	2	1
G	S	S	$0.9 \times 0.1 \times 0.1 = 0.009$	1	2
S	G	G	$0.1 \times 0.9 \times 0.9 = 0.081$	2	1
S	G	S	$0.1 \times 0.9 \times 0.1 = 0.009$	1	2
S	S	G	$0.1 \times 0.1 \times 0.9 = 0.009$	1	2
S	S	S	$0.1 \times 0.1 \times 0.1 = 0.001$	0	3

Putting these results together gives this table:

Good	Substandard	Probability
3	0	0.729
2	1	0.243
1	2	0.027
0	3	0.001

So the probability of more than 1 substandard bulb in a pack is

$0.027 + 0.001 = 0.028$ or 2.8%

This is slightly more than the 2.5% that Samantha regards as acceptable.

What business advice would you give Samantha?

In this example we wrote down all the possible outcomes and found their probabilities one at a time, as you do in a tree diagram. Even with just 3 bulbs this was repetitive. If Samantha had packed her bulbs in boxes of 6, it would have taken 64 lines to list them all. Clearly we need to find a less cumbersome approach.

You will have noticed that in the case of 2 good bulbs and 1 substandard, the probability is the same for each of the three arrangements in the box.

Arrangement			Probability	Good	Substandard
G	G	S	$0.9 \times 0.9 \times 0.1 = 0.081$	2	1
G	S	G	$0.9 \times 0.1 \times 0.9 = 0.081$	2	1
S	G	G	$0.1 \times 0.9 \times 0.9 = 0.081$	2	1

So the probability of this outcome is 3×0.081 (or 0.243). The number 3 arises because there are three ways of arranging two good and one faulty bulb in the box. This is a result you have already met in the previous chapter but written slightly differently.

EXAMPLE

How many different ways are there of arranging the letters *GGS*?

Solution:

There are 3 letters; *G* occurs 2 times and *S* 1 time.

So the number of ways is

$$\overbrace{3 \text{ letters}} \quad \frac{\overset{}{3!}}{\underset{\underbrace{2 \text{ of letter } G}}{2!} \times \underset{\underbrace{1 \text{ of letter } S}}{1!}} = \frac{6}{2} = 3$$

Another way of looking at the situation is to see that since all the letters are either *G* or *S*, all you need to do is to count the number of ways of choosing the letter *G* 2 times out of 3 letters.

EXAMPLE

How many ways are there of selecting 2 objects out of 3?

Solution:

$$^{3}C_2 = \frac{3!}{2! \times 1!} = \frac{6}{2} = 3$$

Clearly it does not matter which way you look at it; the answer is the same.

So what does this tell you? There was no need to list all the possibilities for Samantha's boxes of bulbs. The information could have been written down like this:

Good	Substandard	Expression	Probability
3	0	$^{3}C_3(0.9)^3$	0.729
2	1	$^{3}C_2(0.9)^2\,(0.1)$	0.243
1	2	$^{3}C_1(0.9)^1\,(0.1)^2$	0.027
0	3	$^{3}C_0(0.1^3$	0.001

The Binomial Distribution

Samantha's light-bulbs are an example of a common type of situation which is modelled by the Binomial distribution.

- You are conducting trials on random samples of a certain size, denoted by *n*.

- There are just two possible outcomes (in this case Good and Substandard). These are often referred to as *failure* and *success*.

- Both outcomes have fixed probabilities, the two adding to 1. The probability of success is usually called p, that of failure q, so $p + q = 1$.
- The probability of success in any trial is independent of the outcomes of previous trials.

You can then list the probabilities of the different possible outcomes, as in the table above, or you can write down an expression which summarises the information.

One way of doing this for Samantha's bulbs would be

$$[(0.9G) + (0.1S)]^3$$

where G and S are 'markers' telling you how many Good and Substandard bulbs are being referred to.

This multiplies out to give

$$^3C_0(0.9)^3G^3 + {}^3C_1(0.9)^2(0.1)G^2S + {}^3C_2(0.9)(0.1)^2GS^2 + {}^3C_3(0.1)^3S^3$$
$$\text{or} \quad (0.9)^3G^3 + 3(0.9)^2(0.1)G^2S + 3(0.9)(0.1)^2GS^2 + (0.1)^3S^3$$
$$\text{or} \quad 0.729G^3 + 0.243G^2S + 0.027GS^2 + 0.001S^3$$

In this the numbers 0.729, 0.243, 0.027 and 0.001 refer to the probabilities, and the powers of G and S tell you the numbers of Good and Substandard bulbs. You will notice that the powers of G and S always add up to 3, the total number of bulbs in a box.

If you think about it, you will realise that you do not actually need two markers; one will do. In this case you need only to count the number of faulty bulbs; the rest must be good. This is the way that it is usually done with the letter t used to denote the number of *successes* (whatever you choose to call a success).

In Samantha's case this would then give $[0.9 + 0.1t]^3$

(where success is taken to be finding a faulty bulb), which could be expanded to:

$$0.729 + 0.243t + 0.027t^2 + 0.001t^3$$

In general this would be written

$$[q + pt]^n$$

where p is the probability of a success and q is the probability of a failure.

An expression like this is called a *probability generating function*. The Binomial distribution for a sample of size n with probability of success p is denoted by $B(n, p)$. So the distribution of the probability of different numbers of substandard bulbs in Samantha's boxes is $B(3, 0.1)$.

In general the probability that the number of successes, X, in n trials is r is given by

$$P(X = r) = {}^nC_r p^r [1 - p]^{n-r} = {}^nC_r p^r q^{n-r} \qquad 0 \le r \le n$$

The reason that this sort of distribution is called binomial is because the probability generating function is a binomial expression.

The number of successes, X, in n trials is an example of a random variable. The value of X varies in a random manner from one set of trials to another. The names of random variables are denoted by upper case letters, like X, but the particular values they take are given by lower case letters, like r.

The Binomial Theorem

You may well have covered the work in this section in your Pure Mathematics.

A typical binomial expression may be written

$$(x + y)^n.$$

'Binomial' means 'two numbers' and these are x and y. When the binomial expression is a probability generating function, n is a positive whole number and x and y represent probabilities adding to 1.

There are two common ways of expanding, or multiplying out, a binomial expression, by using Pascal's triangle or the formula for ^nC_r.

Pascal's Triangle

If you multiply out $(x + y)^4$, you get $1x^4 + 4x^3y + 6x^2y^2 + 4xy^3 + 1y^4$. You can find the various coefficients in this expression (1, 4, 6, 4, 1) by looking along row 4 of Pascal's Triangle, below.

```
Row
0                            1
1                        1       1
2                    1       2       1
3                1       3       3       1
4            1       4       6       4       1
5        1       5      10      10      5       1
                    and so on
```

To find the next row of Pascal's Triangle you put 1 in both of the outside positions, and then find the number to go in each place by adding the two diagonally above it.

The numbers in row 5 are

$$1 \quad 5 \quad 10 \quad 10 \quad 5 \quad 1$$

so row 6 is

$$1 \quad 6 \quad 15 \quad 20 \quad 15 \quad 6 \quad 1$$

Pascal's Triangle is a quick and easy way of working out the Binomial coefficients if n is not too large. However it soon becomes cumbersome, particularly if you only want to work out a single coefficient.

The other method is to use the formula

$$^nC_r = \frac{n!}{r! \times (n-r)!}.$$

You can check this for yourself, by taking, say, $n = 6$. In that case the coefficients

$$^6C_0 \; ^6C_1 \; ^6C_2 \; ^6C_3 \; ^6C_4 \; ^6C_5 \; ^6C_6$$

should be the same as row 6 of Pascal's Triangle:

$$1 \quad 6 \quad 15 \quad 20 \quad 15 \quad 6 \quad 1$$

Remember that by convention $0! = 1$.

By looking at Pascal's Triangle, you can see two useful results concerning Binomial coefficients:

1. If you add the coefficients in each row you will find that the answers are:

Row	0	1	2	3	4	5	6
Total	1	2	4	8	16	32	64

which leads you to suspect the general result that

$$\sum_{r=0}^{n} {}^nC_r = 2^n.$$

Actually that result is obvious if you put $x = 1$ and $y = 1$ in the expansion.

2. You have seen how to add the coefficients on one line of Pascal's Triangle to get those on the next. You can write this as a general rule:

$$^nC_r + {}^nC_{r+1} = {}^{n+1}C_{r+1}.$$

Using Your Tables or Calculator

You can also find binomial coefficients by looking them up in your tables, or by using the ^nC_r function on your calculator.

TABLE OF BINOMIAL COEFFICIENTS $\binom{n}{r} = {}^nC_r$

$r = 0$	1	2	3	4	5	6	7	8	9	10
$n =$ 1	1	1								
2	1	2	1							
3	1	3	3	1						
4	1	4	6	4	1					
5	1	5	10	10	5	1				
6	1	6	15	20	15	6	1			
7	1	7	21	35	35	21	7	1		
8	1	8	28	56	70	56	28	8	1	
9	1	9	36	84	126	126	84	36	9	1
10	1	10	45	120	210	252	210	120	45	10	1
11	1	11	55	165	330	462	462	330	165	55	11
12	1	12	66	220	495	792	924	792	495	220	66
13	1	13	78	286	715	1287	1716	1716	1287	715	286
14	1	14	91	364	1001	2002	3003	3432	3003	2002	1001
15	1	15	105	455	1365	3003	5005	6435	6435	5005	3003
16	1	16	120	560	1820	4368	8008	11440	12870	11440	8008
17	1	17	136	680	2380	6188	12376	19448	24310	24310	19448
18	1	18	153	816	3060	8568	18564	31824	43758	48620	43758
19	1	19	171	969	3876	11628	27132	50388	75582	92378	92378
20	1	20	190	1140	4845	15504	38760	77520	125970	167960	184756

Exercise 4A

Write out these binomial expansions in full:

1. $(a + b)^2$;   2. $(c + d)^3$;   3. $(e + f)^4$;

4. $(l + m)^5$;   5. $(p + q)^6$;   6. $(s + t)^8$;

7. $(h/2 + t/2)^4$;   8. $(5n/6 + s/6)^5$;   9. $(6/7 + t/7)^6$;

10. $(3/4 + t/4)^7$

Find the terms requested in these binomial expansions:

11. The term in $h^2 t^2$ in $(h/2 + t/2)^4$

12. The term in $n^2 y^3$ in $(2n/3 + y/3)^5$

13. The term in $n^4 s^2$ in $(5n/6 + s/6)^6$

14. The term in t^3 in $(6/7 + t/7)^{15}$

15. The term in t^{15} in $(3/4 + t/4)^{25}$

Using the Binomial Distribution

THE AVONFORD STAR

Calvin's Tips

If you are to be successful at the tables, you must develop an insight into probability. No one makes money playing against the odds in the long term. Try to argue your way round this problem.

Which is more likely: that you get at least 1 six when you throw a die 6 times, or that you get at least 2 sixes when you throw it 12 times?

On a single throw of a die the probability of getting a six is 1/6 and that of not getting a six is 5/6.

So the probability generating functions for the two situations in Calvin's problem are:

$(5/6 + 1/6t)^6$ and $(5/6 + 1/6t)^{12}$

giving probabilities of:

$1 - {}^6C_0(5/6)^6 = 1 - 0.335 = 0.665$ (at least 1 six in 6 throws)

and

$1 - ({}^{12}C_0(5/6)^{12} + {}^{12}C_1(5/6)^{11}(1/6)) = 1 - (0.112 + 0.269)$
$= 0.619$ (at least 2 sixes in 12 throws)

So at least 1 six in 6 throws is somewhat more likely.

EXAMPLE

Extensive research has shown that 1 person out of every 4 is allergic to a particular grass seed. A group of 20 university students volunteer to try out a new treatment.
(i)   What is the expectation of the number of allergic people in the group?
(ii) What is the probability that (A) exactly 2, (B) no more than 2, of the group are allergic?
(iii) How large a sample would be needed for the probability of it containing at least one allergic person to be greater than 99.9%.
(iv) What assumptions have you made in your answer?

Solution:
In this situation $n = 20$, $p = \frac{1}{4}$ and $q = \frac{3}{4}$. The number of allergic people is denoted by X.
(i) Expectation $= np = 20 \times \frac{1}{4} = 5$ people.

(ii) The probability generating function is

$$[\tfrac{3}{4} + \tfrac{1}{4}t]^{20}$$

(A) $P(X = 2) = {}^{20}C_2(\tfrac{3}{4})^{18} (\tfrac{1}{4})^2 = 0.067$

(B) $P(X \le 2) = P(0) + P(1) + P(2)$

$$= (\tfrac{3}{4})^{20} + {}^{20}C_1 (\tfrac{3}{4})^{19} (\tfrac{1}{4}) + {}^{20}C_2(\tfrac{3}{4})^{18} (\tfrac{1}{4})^2$$

$$= 0.003 + 0.021 + 0.067$$

$$= 0.091$$

(iii) Let the sample size be n (people)

The probability that none of them is allergic is

$$P(X = 0) = (\tfrac{3}{4})^n$$

and so the probability that at least 1 is allergic is

$$P(X \ge 1) = 1 - P(X = 0)$$
$$= 1 - (\tfrac{3}{4})^n$$

So we need
$$1 - (0.75)^n > 0.999$$
$$(0.75)^n < 0.001$$

Solving
$$(0.75)^n = 0.001$$

gives
$$n \log 0.75 = \log 0.001$$
$$n = \log 0.001 \div \log 0.75$$
$$= 24.01$$

So 25 people are required.

N O T E   *Although 24.01 is very close to 24 it would be incorrect to round down.*
$1 - (0.75)^{24} = 0.9989966$ which is just less than 99.9%.

(iv) The assumptions made are:

(a) That the sample is random. This is almost certainly untrue. University students are nearly all in the 18–25 age range, and they may well be unrepresentative of the whole population in other ways. Volunteers are seldom truly random.

(b) That the outcome for one person is independent of that for another. This is probably true, unless they were a group of friends from, say, an athletics team, where those with allergies were less likely to be members.

Does the Binomial Distribution Really Work?

Experiment
In Calvin's first case you threw a die 6 times (or six dice once each, which amounts to the same thing). The probability generating function for this is

$$(5/6 + 1/6t)^6$$

and this gives the probabilities in the following table.

Experiment continued

Number of sixes	Probability
0	0.335
1	0.402
2	0.201
3	0.054
4	0.008
5	0.001
6	0.000

So if you carry out the experiment of throwing six dice 1000times and record the number of sixes each time, you should get none about 335 times, one about 402 times and so on. What does 'about' mean? How close agreement can you expect between experimental and theoretical results?

You could carry out the experiment with dice, but it would be very tedious even if several people each shared the work. Alternatively you could simulate the experiment on a computer, using its random number generator.

This program, written in BBC BASIC, allows you to conduct this experiment.

```
10 DIM S(7)
20 FOR I=1 TO 7
30    S(I)=0
40 NEXT I
50 INPUT "How many throws of the six dice ";N
60 FOR T=1 TO N
70    X=0
80    FOR D=1 TO 6
90       Y=RND(6)
100       IF Y=6 THEN X=X+1
110    NEXT D
120    S(X+1)=S(X+1)+1
130 NEXT T
140 FOR J=0 TO 6
150    PRINT J,S(J+1)
160 NEXT J
170 END
```

Investigation

(This is a true story.)
During voting at a by-election, an exit poll of 1700 voters indicated that 50% of people had voted for the Labour party candidate. When the real votes were counted it was found that he had in fact received 57% support.

Carry out this computer simulation of the situation and use it to decide whether the difference was likely to have occurred because of the random nature of the sample, or because of faulty sampling techniques or for other possible reasons you can think of.

```
10 INPUT "How many samples of size 1700 ";N
20 FOR S=1 TO N
30    L=0
40    FOR V=1 TO 1700
50       X=RND(100)
60       IF X<=57 THEN L=L+1
70    NEXT V
80    LET P=L*100/1700
90    LET P=INT(P*10+0.5)/10
100   PRINT P
110 NEXT S
120 END
```

Exercise 4B

1. The recovery ward in a maternity hospital has 6 beds. What is the probability that the mothers there have between them 4 girls and 2 boys? (You may assume that there are no twins, and that a baby is equally likely to be a girl or a boy.)

2. A typist has a probability of 0.99 of typing a letter correctly. He makes his mistakes at random. He types a sentence containing 200 letters. What is the probability that he makes exactly one mistake?

3. In a well-known game you have to decide which your opponent is going to choose: 'Paper', 'Stone' or 'Scissors'. If you guess entirely at random, what is the probability that you are right exactly 5 times out of 15?

4. There is a fault in a machine making microchips, with the result that only 80% of those it produces work. A random sample of 8 microchips made by this machine is taken. What is the probability that exactly 6 of them work?

5. An airport is situated in a place where poor visibility (less than $800\,$m) can be expected 25% of the time. A pilot flies into the airport on 10 different occasions.
   (i) What is the probability that he encounters poor visibility exactly 4 times?
   (ii) What other factors could influence the probability?

6. Three coins are tossed.
   (i)   What is the probability of all three showing heads?
   (ii) What is the probability of two heads and one tail?
   (iii) What is the probability of one head and two tails?
   (iv) What is the probability of all three showing tails?
   (v)   Show that the probabilities for the four possible outcomes add up to 1.

S1

Exercise 4B continued

7. A coin is tossed ten times.
   (i) What is the probability of it coming down heads 5 times and tails 5 times?
   (ii) Which is more likely: exactly 7 heads or more than 7 heads?

8. In an election 30% of people support the Progressive Party. A random sample of 8 voters is taken. What is the probability that it contains:
   (i) 0; (ii) 1; (iii) 2; (iv) at least 3 supporters of the Progressive Party?
   (v) Which is the most likely number of Progressive Party supporters to find in a sample of size 8?

9. There are 15 children in a class. What is the probability that:
   (i) 0; (ii) 1; (iii) 2; (iv) at least 3
   were born in January? What assumption have you made in answering this question? How valid is that assumption in your view?

10. Criticise this argument
    If you toss two coins they can come down three ways: two heads, one head and one tail, or two tails. There are three outcomes and so each of them must have probability one third.

11. In a game 5 dice are rolled together. What is the probability that
    (i) all 5 show 1; (ii) exactly 3 show 1; (iii) none of them shows 1.
    (iv) What is the most likely number of times for 6 to show?

12. (i) What is the probability that exactly 2 children in a class of 10 were born on a Monday?
    (ii) Which is the more likely for the same class of 10 children: that fewer than 2 children were born on a Monday or that more than 2 of them were?

13. There are eight colours of Smarties which normally occur in equal proportions: red, orange, yellow, green, blue, purple, pink and brown. Veronica's mother gives each of her children 16 Smarties. Veronica says that the blue ones are much nicer than the rest and is very upset when she receives less than her fair share of them.
    (i) How many blue Smarties did Veronica expect to get?
    (ii) What was the probability that she would receive fewer blue ones than she expected?
    (iii) What was the probability that she would receive more blue ones than she expected?

14. In a particular area, 30% of men and 20% of women are overweight, and there are 4 men and 3 women working in an office there. Find the probability that there are

Exercise 4B continued

(i) 0; (ii) 1; (iii) 2 overweight men;
(iv) 0; (v) 1; (vi) 2 overweight women;
(vii) 2 overweight people in the office.
What assumption have you made in answering this question?

15. Akosua's class take a multiple choice test with 10 questions, with the answer to each of them having 4 options, *A*, *B*, *C* and *D*. Akosua gets all 10 questions right but the rest of the class tell her it was a fluke and she must have guessed them all. Akosua replies that the chance of doing that is less than one in a million.
    (i)   Is Akosua's reply correct?
    (ii) What is the expectation of the number of correct answers for someone who guesses all 10 answers?
    (iii) Is someone who guesses all 10 answers more likely to get 2 or 3 of them right?

16. On her drive to work Stella has to go through 4 sets of traffic lights. She estimates that for each set the probability of her finding them red is ⅔ and green ⅓. (She ignores the possibility of their being amber.) Stella also estimates that when a set of lights is red she is delayed by 1 minute. Find the probability of
    (i) 0; (ii) 1; (iii) 2; (iv) 3 sets of lights being against her.
    (v) Find the expected extra journey time due to waiting at lights.

17. A drunk man steps out of a bar into a narrow alley which runs from West to East. At each step he chooses at random whether to go East or West. After 12 steps he stops for a rest.
    (i)   Why is it impossible for him then to be 1 step away from the bar?
    (ii) What is the probability that he is then 10 steps East of the bar?
    (iii) What is his most likely position?
    (iv) What is the probability that he is 4 steps away from the bar, either to the East or to the West?
    (v)   What is the expectation of his number of steps from the bar (ignoring which side he is on, as in part (iv)).

18. Pepper moths are found in two varieties, light and dark. The proportion of dark moths increases with certain types of atmospheric pollution. At the time of the question 30% of the moths in a particular town are dark. A research student sets a moth trap and catches 9 moths, 4 light and 5 dark.
    (i) What is the probability of that result for a sample of 9 moths?
    (ii) What is the expected number of dark moths in a sample of 9?
    The next night the student's trap catches 10 pepper moths.
    (iii) What is the probability that the number of dark moths in this sample is the same as the expected number?

Exercise 4B continued

19. An insurance salesman sells policies to 5 men, all of identical age and in good health. According to his company's records the probability that a man of this particular age will be alive in 20 years' time is ⅔. Find the probability that in 20 years' time the number of men still alive will be
(i) five, (ii) at least three, (iii) exactly two, (iv) at least one.

[（

20. Five fair dice are thrown. Find the probabilities that there are
(i)   exactly 3 sixes; (ii) at least 3 sixes;
(iii) 3 sixes and 2 fives;
(iv) 2 fives, given that there are exactly 3 sixes showing.

[ME

21. Bella Cicciona, a fortune teller, claims to be able to predict the sex o
unborn children. In fact, on each occasion she is consulted, the probability that she makes a correct prediction is 0.6, independent o
any other prediction.
One afternoon, Bella is consulted by 10 expectant mothers. Find, correct to 2 significant figures, the probabilities that
(i)   her first 8 predictions are correct and her last 2 are wrong;
(ii) she makes exactly 8 correct predictions;
(iii) she makes at least 8 correct predictions;
(iv) she makes exactly 8 correct predictions given that she makes at least 8.

[ME

22. A general knowledge quiz has 10 questions. Each question has 3 possible 'answers' of which 1 only is correct. A woman attempts the quiz by pure guesswork. Find the probabilities that she obtains
(i) exactly 2 correct answers,   (ii) not more than 2 correct answers.
What is the most likely number of correct answers and the probability that she just achieves this number?

[ME

23. A certain strain of rabbit produces offspring in equal ratio male to female. One third of the male offspring and one half of the female offspring are black. In a litter of 5 offspring, what is the probability that
(i)   exactly 2 offspring will be male;
(ii) less than 4 offspring will be female?
In a litter of 4 offspring, what is the probability that there will be exactly 2 male offspring which are not black?

[ME

24. Five unbiased dice are thrown. Calculate the probabilities that
(i)   there will be exactly four sixes;
(ii) there will be some one number appearing exactly four times;
(iii) there will be some one number appearing exactly three times and a second number appearing twice.

[ME

25. A company uses machines to manufacture wine glasses. Because of imperfections in the glass it is normal for 10% of the glasses to leave the machine cracked. The company takes regular samples of size 10 from each machine. If more than 2 glasses in a sample are cracked, they stop the machine and check that it is set correctly.
What is the probability that a sample of size 10 contains
   (i)   0; (ii) 1; (iii) 2 faulty glasses, when the machine is correctly set?
   (iv) What is the probability that as a result of taking a sample a machine is stopped when it is correctly set?
   (v)   A machine is in fact incorrectly set and 20% of the glasses from it are cracked. What is the probability that this is undetected by a particular sample?

KEY POINTS

The binomial distribution may be used to model situations in which:

- you are conducting trials on random samples of a certain size, n;
- there are two possible outcomes, often referred to as success and failure;
- both outcomes have fixed probabilities, p and q;
- the probability of success in any trial is independent of the outcomes of previous trials.
- The probability that the number of successes, X, has the value r, is given by the coefficient of t^r in the expansion of

$$(q + pt)^n$$

and so $$P(X = r) = {}^nC_r q^{n-r} p^r.$$

$${}^nC_r = \binom{n}{r} = \frac{n!}{r!(n-r)!} : \quad (0! = 1)$$

- Values of binomial coefficients may also be found using Pascal's Triangle, or using tables.

To be and not to be, that is the answer. $(X = r : r > 0, r < n)$
                                                   Piet Hein

5

You may prove anything by figures.

An anonymous salesman

Machoman Dan

Beer drinking Dan Ellis has just become a father for the eighth time, and they are ALL BOYS.

Find Dan at his local and he will tell you that having no girls among the eight shows that he is a real man.

"When I was just a lad at school I said I had got macho chromo-somes," Dan claims.

"Now I have been proved right."

Cheers! Dan Ellis reckons his all-boy offspring prove he's a real man

What do you think?

There are two quite different points here.

The first is that you have probably decided that Dan is a male chauvinist, preferring boys to girls. However, you should not let your views on that influence your judgement on the second point, his claim to be biologically different from other people, with special chromosomes.

There are two ways this claim could be investigated, to look at his chromosomes under a high magnification microscope or to consider the statistical evidence. Since you have neither Dan Ellis nor a suitable microscope to hand, you must resort to the latter.

If you have eight children you would expect them to be divided about evenly between the sexes, 4 – 4, 5 – 3 or perhaps 6 – 2. When you realised that number nine was on its way you would think it equally likely to be a boy or a girl until it were born, or a scan were carried out, when you would know for certain one way or the other.

In other words you would say that the probability of its being a boy was ½ and that of its being a girl ½. So you can model the number of boys among eight children by the binomial distribution B(8, ½) generated by

$$(0.5 + 0.5t)^8$$

The marker t counts the number of boys among the eight. This can be expanded to

$$^8C_0(0.5)^8 + {}^8C_1(0.5)^7\,(0.5t) + {}^8C_2(0.5)^6\,(0.5t)^2 + \ldots + {}^8C_7(0.5)\,(0.5t)^7$$
$$+ {}^8C_8(0.5t)^8$$

This gives these probabilities, shown in figure 5.1

Boys	Girls	Probability
0	8	1/256
1	7	8/256
2	6	28/256
3	5	56/256
4	4	70/256
5	3	56/256
6	2	28/256
7	1	8/256
8	0	1/256

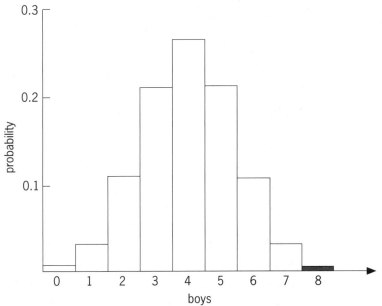

Figure 5.1

So you can say that if a biologically normal man fathers eight children, the probability that they will all be boys is 1/256.

This is unlikely but by no means impossible.

N O T E   *The probability of a baby being a boy is not in fact 0.5 but about 0.503. Boys are less tough than girls and so more likely to die in infancy and this seems to be nature's way of compensating. In most societies men have a markedly lower life expectancy as well.*

For Discussion: A Modelling Investigation

You may think Dan Ellis a thoroughly objectionable character, but in some countries large sections of society value boy children more highly than girls. Medical advances mean that it will soon be possible to decide in advance the sex of your next baby. What would be the effect of this on a country's population if, say, half the parents decided to have only boys and the other half to let nature take its course?

(This is a real problem. The social consequences could be devastating.)

Defining Terms

In the last example we investigated Dan Ellis's claim by comparing it to the usual situation, the unexceptional. If we use p for the probability that a child is a boy then the normal state of affairs can be stated as

$$p = 0.5$$

This is called the *Null hypothesis*, denoted by H_0.

Dan's claim (made, he says, before he had any children) was that

$$p > 0.5$$

and this is called the *Alternative hypothesis*, H_1.

The word *hypothesis* (plural *hypotheses*) means a theory which is put forward either for the sake of argument or because it is believed or suspected to be true. An investigation like this is usually conducted in the form of a test, called a *Hypothesis test*. There are many different sorts of hypothesis test used in Statistics; in this chapter you meet only one of them.

It is never possible to prove something statistically in the sense that, for example, you can prove that the angle sum of a triangle is 180°. Even if you tossed a coin a million times and it came down heads every single time, it is still possible that the coin is unbiased and just happened to land that way. What you can say is that it is very unlikely; the probability of it happening that way is $(\frac{1}{2})^{1\,000\,000}$ which is a decimal that starts with over 300 000 zeros. This is so tiny that you would feel quite confident in declaring the coin biased.

There comes a point when the probability is so small that you say "That's good enough for me. I am satisfied that it hasn't happened that way by chance". The probability at which you make that decision is called the *Significance level* of the test. Significance levels are usually given as percentages; 0.05 is written 5%, 0.01 as 1% and so on.

So in the case of Dan Ellis, the question could have been worded:

Test at the 1% significance level, Dan Ellis's boyhood claim that his children are more likely to be boys than girls.

The answer would then look like this:

Null hypothesis: H_0 $p = 0.5$ (boys and girls are equally likely)
Alternative hypothesis: H_1 $p > 0.5$ (boys are more likely)
Significance level: 1%

Probability of 8 boys from 8 children $= 1/256 = .0039 = 0.39\%$

Since $0.39\% < 1\%$ we reject the null hypothesis and accept the alternative hypothesis.

We accept Dan Ellis's claim.

This example also illustrates some of the problems associated with hypothesis testing. Here is a list of points you should be considering.

Hypothesis Testing Checklist

1 *Was the test set up before or after the data were known?*

The test consists of a null hypothesis, an alternative hypothesis, and a significance level.

In this case, the null hypothesis is the natural state of affairs and so does not really need to be stated in advance. Dan's claim "When I was just a lad at school I said I had got macho chromosomes" could be interpreted as the Alternative hypothesis, $p > 0.5$.

The problem is that one suspects that whatever children Dan had he would find an excuse to boast. If they had all been girls, he might have been talking about "my irresistible attraction for the opposite sex" and if they had been a mixture of girls and boys he would have been claiming "super-virility".

Any test carried out retrospectively must be treated with suspicion.

2 *Is the data used for the test chosen independently and at random?*

If there are no twins, triplets etc among Dan's children, then they are independent, but certainly not random if they all had the same mother.

3 *Is the statistical procedure actually testing the original claim?*

Dan Ellis claims to have 'macho chromosomes' whereas the statistical test is of the alternative hypothesis that $p > 0.5$. The two are not necessarily the same. Even if this alternative hypothesis is true, it does not necessarily follow that Dan has macho chromosomes.

The Ideal Hypothesis Test

In the ideal hypothesis test, you take the following steps, in this order:

(a) *Establish the null and alternative hypotheses.*
(b) *Decide on the significance level.*
(c) *Collect suitable data, using a sampling procedure that ensures the items are independent and selected at random.*
(d) *Conduct the test, doing the necessary calculations.*
(e) *Interpret the result in terms of the original claim, theory or problem.*

There are times however when you need to carry out a test but it is just not possible to do so as rigorously as this.

In the case of Dan Ellis, you would require him to go away and father eight more children, this time with randomly selected mothers, which is clearly impossible. Had Dan been a laboratory rat and not a human, however, you probably could have organised it.

Choosing the Significance Level

If, instead of 1%, we had set the significance level at 0.1%, then we would have rejected Dan's claim, since 0.39% > 0.1%. The lower the percentage in the significance level, the more stringent is the test.

The significance level you choose for a test involves a balanced judgement.

Imagine that you are testing the rivets on an aeroplane's wing to see if they have lost their strength. Setting a small significance level, say 0.1%, means that you will only declare the rivets weak if you are very confident of your finding. The trouble with requiring such a high level of evidence is that even when they are weak you may well fail to register the fact, with the consequence that the aeroplane crashes. On the other hand if you set a high significance level, like 10%, you run the risk of declaring the rivets faulty when they are all right, involving the company in expensive and unnecessary maintenance work.

The question of how you choose the best significance level is, however, beyond the scope of this introductory chapter.

EXAMPLE

Here is another example, from 'Calvin's Tips' in *The Avonford Star.* Cover up the solution and then, as you work your way through it, see if you can predict the next step at each stage.

Does the procedure follow the steps of the ideal hypothesis test?

Calvin's Tips

I was at a casino the other evening. After several hours play, a lady whom I shall call Leonora had lost a lot of money. She complained to the management that one of the dice was biased, with a tendency to show the number 1.

The management agreed to test the die at the 5% Significance Level, throwing it 20 times. If the test supported Leonora's view she would get her money refunded, otherwise she would be asked to leave the premises and never return.

The results were as follows:

1	6	6	5	5
1	2	3	2	3
4	4	4	1	4
1	1	4	1	3

What happened to my friend Leonora?

Solution

Let p be the probability of getting 1 on any throw of the die.

Null hypothesis: H_0 $p = 1/6$ (The die is unbiased)
Alternative hypothesis: H_1 $p > 1/6$ (The die is biased towards 1)
Significance level: 5%

The results may be summarised as follows:

Score   1 2 3 4 5 6
Frequency   6 2 3 5 2 2

The situation is modelled by the Binomial distribution, B(20, $1/6$), generated by

$$[5/6 + 1/6 t]^{20}$$

where the marker t stands for success (getting 1).

Expanding this gives these probabilities:

Number of successes	Expression	Probability
0	$(5/6)^{20}$	0.0261
1	$^{20}C_1 (5/6)^{19}(1/6)$	0.1043
2	$^{20}C_2 (5/6)^{18}(1/6)^2$	0.1982
3	$^{20}C_3 (5/6)^{17}(1/6)^3$	0.2379
4	$^{20}C_4 (5/6)^{16}(1/6)^4$	0.2022
5	$^{20}C_5 (5/6)^{15}(1/6)^5$	0.1294
	Total	0.8981

So, calling X the number of ones occuring when a die is rolled 20 times, the probability of 6 or more ones is given by $P(X \geq 6) = 1 - P(X \leq 5) = 1 - 0.8981 = 0.1019$, about 10%.

Since 10% > 5%, the null hypothesis (the die is unbiased) is accepted.

The alternative hypothesis (the die is biased in favour of the number 1) is rejected, even though the 1 did come up more often than the other numbers.

> **THE AVONFORD STAR**
>
> Leonora did not get her money back but the management relented and did not ban her from the Casino.

NOTE

Notice that this is a test not of the particular result (6 ones) but of a result at least as extreme as this (at least 6 ones), the area shaded in figure 5.2.

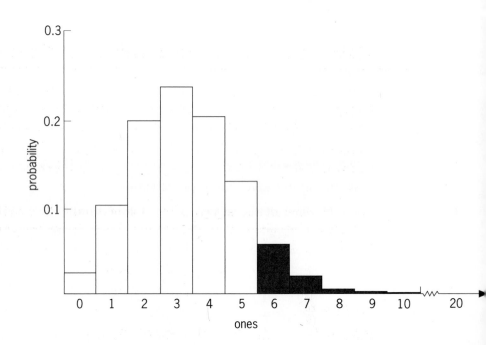

Figure 5.2

Another Method

You could have found the probability of up to 5 scores of 1 using *Cumulative Binomial Probability* tables.

Look up under $n = 20$

n	x	p 0.050	0.100	0.150	1/6	0.200	0.250	0.300	1/3	0.350
20	0	0.3585	0.1216	0.0388	0.0261	0.0115	0.0032	0.0008	0.0003	0.0002
	1	0.7358	0.3917	0.1756	0.1304	0.0692	0.0243	0.0076	0.0033	0.0021
	2	0.9245	0.6769	0.4049	0.3287	0.2061	0.0913	0.0355	0.0176	0.0121
	3	0.9841	0.8670	0.6477	0.5665	0.4114	0.2252	0.1071	0.0604	0.0444
	4	0.9974	0.9568	0.8298	0.7687	0.6296	0.4148	0.2375	0.1515	0.1182
	5	0.9997	0.9887	0.9327	0.8982	0.8042	0.6172	0.4164	0.2972	0.2454
	6	1.0000	0.9976	0.9781	0.9629	0.9133	0.7858	0.6080	0.4793	0.4166
	7		0.9996	0.9941	0.9887	0.9679	0.8982	0.7723	0.6615	0.6010
	8		0.9999	0.9987	0.9972	0.9900	0.9591	0.8867	0.8095	0.7624
	9		1.0000	0.9998	0.9994	0.9974	0.9861	0.9520	0.9081	0.8782
	10			1.0000	0.9999	0.9994	0.9961	0.9829	0.9624	0.9468
	11				1.0000	0.9999	0.9991	0.9949	0.9870	0.9804
	12					1.0000	0.9998	0.9987	0.9963	0.9940
	13						1.0000	0.9997	0.9991	0.9985
	14							1.0000	0.9998	0.9997
	15								1.0000	1.0000

In this case $p = \frac{1}{6}$ so the probability of up to 5 scores of 1 is 0.8982, the same result as before apart from the last figure where there is a difference of 1 from rounding.

Critical Value and Critical Region

In the above example the number 1 came up six times and this was not enough to get poor Leonora a refund. What was the least number of times 1 would have had to come for the test to give the opposite result?

As on the previous page, we use X to denote the number of times one comes up in the 20 throws, and so $X = 6$ means that the number one comes up 6 times.

We know from our earlier work that the probability that $X \leq 5$ is 0.8982 and we can use the Binomial distribution to work out the probabilities that $X = 6$, $X = 7$, etc.

$$P(X = 6) = {}^{20}C_6 \, (\tfrac{5}{6})^{14} \, (\tfrac{1}{6})^6 = 0.0647$$
$$P(X = 7) = {}^{20}C_7 \, (\tfrac{5}{6})^{13} \, (\tfrac{1}{6})^7 = 0.0225$$

We know

$$P(X \geq 6) = 1 - P(X \leq 5) = 1 - 0.8982 = 0.1018$$

0.1018 is a little over 10% and so greater than the significance level of 5%. There is no reason to reject H_0.

What about the case when the number one comes up 6 times, that is $X = 6$?

$$\text{Since } P(X \leq 6) = P(X \leq 5) + P(X = 6)$$
$$P(X \leq 6) = 0.0892 + 0.0647 = 0.9629$$

$$\text{So } P(X \geq 7) = 1 - P(X \leq 6)$$
$$= 1 - 0.9629 = 0.0371 = 3.71\%$$

Since $3.7\% < 5\%$ H_0 is now rejected in favour of H_1.

We have got there. You can see that Leonora needed the 1 to come up 7 or more times if her case was to be upheld. She missed by just one.

In this example the number 7 is the *Critical value* (at the 5% significance level), the value at which you change from accepting the null hypothesis to rejecting it. The range of values for which you reject the null hypothesis, in this case $X \geq 7$ is called the *Critical region*.

It is sometimes easier in hypothesis testing to find the critical region and see if your value lies in it, rather than working out the probability of a value at least as extreme as the one you have, the procedure used so far. This is done in the last example in this chapter on page 133.

Do you think Leonora's 'all or nothing' test was a bit harsh? Sometimes tests are designed so that if the result falls within a certain region further trials are recommended.

Experiments

Mind Reading
Here is a simple experiment to see if you can read the mind of a friend whom you know well. The two of you face each other across a table on which is placed a coin. Your friend takes the coin, and puts it in one or other hand under the table. You have to guess which one.

Play this game at least 20 times and test at the 10% Significance level whether you can read your friend's mind.

Smarties
Get a large box of Smarties and taste the different colours. Choose the colour, C, which you think has the most distinctive flavour.

Now close your eyes and get a friend to feed you Smarties. Taste each one and say if it is your chosen colour or not. Do this for at least 20 Smarties and test at the 10% significance level whether you can pick out those with colour C by taste.

Left and Right

It is said that if people are following a route which brings them to a T-junction where they have a free choice between turning left and right, the majority will turn right.

Design and carry out an experiment to test this hypothesis.

N O T E *This is taken very seriously by companies choosing stands at exhibitions. It is considered worth paying extra for a location immediately to the right of one of the entrances.*

Exercise 5A

In all these questions you should apply the checklist on page 121 to the hypothesis test:

Was the test set up before or after the data were known?
Are the data used for the test chosen independently and at random?
Is the statistical procedure actually testing the original claim?

You should comment critically on whether the steps on page 121 have been followed:

(a) *Establish the null and alternative hypotheses.*
(b) *Decide on the significance level.*
(c) *Collect suitable data, using a sampling procedure that ensures the items are independent and selected at random.*
(d) *Conduct the test, doing the necessary calculations.*
(e) *Interpret the result in terms of the original claim, theory or problem.*

1. Mrs da Silva is running for President. She claims to have 60% of the population supporting her.

   She is suspected of overestimating her support and a random sample of 20 people are asked whom they support. Only 9 say Mrs da Silva.

   Test, at the 5% Significance level, the hypothesis that she has overestimated her support.

2. 60% of the pupils in a school are girls. The head teacher claims that the girls work harder than the boys.

   There are 20 classes in the school and at the end of each year a pupil is chosen from each class, by the class tutor, for a prize for hard work. The year after the head's claim, 17 of the prizewinners are girls.

   Does the evidence support the head's claim at the 10% significance level?

3. A driving instructor claims that 60% of his pupils pass their driving test at the first attempt. Supposing this claim is true, find, using tables or otherwise, the probability that, of 20 pupils taking their first test,

   (i) exactly 12 pass,

   (ii) more than 12 pass.

   A local newspaper reporter suspects that the driving instructor is exaggerating his success rate, and so she decides to carry out a statistical investigation. State the null and alternative hypotheses which she is testing. She contacts 20 of his pupils who have recently taken their first test and discovers that N passed. Given that she performs a 5% significance test and that she concludes that the driving instructor's claim was exaggerated, what are the possible values of N?

   [MEI]

4. World wide 25% of men are colourblind but it is believed that the condition is less widespread among a group of remote hill tribes. An anthropologist visits a village in that area and finds that only 2 out of 30 men he tests are colourblind.

   Does this evidence support, at the 5% significance level, the belief that fewer men than normal are colourblind in these hill tribes?

5. During the 12 years 1980 to 1991 Oxford University won the Boat Race eleven times, Cambridge only once in 1986.

   After the 1983 race, a commentator stated that university policies now made it more likely that Oxford would win than Cambridge.

   Carry out a hypothesis test at the 10% significance level and state what conclusions can be drawn from it.

6. A company developed synthetic coffee and claim that coffee drinkers could not distinguish it from the real product. A number of coffee drinkers refuted the company's claim, saying that the synthetic coffee tasted synthetic. In a test, carried out by an independent consumer protection body, 20 people were given a mug of coffee. Ten had the synthetic brand and ten the natural, but they were not told which they had been given.

   Out of the 10 given the synthetic brand, 8 said it was synthetic and 2 said it was natural. Use this information to test the coffee drinkers' claim (as against the null hypothesis of the company's claim), at the 5% significance level.

7. A sceptic challenges a fortune teller who believes that the saying *Monday's child is fair of face* is true.

   At a local beauty competition the next month there are 20 entrants. Their birthdays are given on the programme and from them the sceptic works out that 7 were born on a Monday.

Exercise 5A continued

State suitable null and alternative hypotheses and test the conflicting claims at the 5% significance level.

8. A group of 18 students decides to investigate the truth of the saying that if you drop a piece of toast it is more likely to land butter side down. They each take one piece of toast, butter it on one side and throw it in the air. Eleven land butter side down, the rest butter side up. Use their results to carry out a hypothesis test at the 10% significance level, stating clearly your null and alternative hypotheses.

9. On average 70% of people pass their driving test first time. There are complaints that Mr McTaggart is too harsh and so, unknown to himself, his work is monitored. It is found that he fails 10 out of 20 candidates. Are the complaints justified at the 5% significance level?

10. A machine makes bottles. In normal running 5% of the bottles are expected to be cracked, but if the machine needs servicing this proportion will increase. As part of a routine check, 50 of its bottles are inspected and 5 are found to be unsatisfactory. Does this provide evidence, at the 5% significance level, that the machine needs servicing?

11. A long-life light bulb is designed so that 80% of them have lifetimes of at least 10 000 hours. Mrs Smith believes this claim to be exaggerated. She buys 20 of them and monitors carefully the time each of them is turned on. She finds that 7 of them have lifetimes less than the advertised 10 000 hours and complains to the manufacturer that their claim is false. Is her complaint justified?

12. A customer complains to a casino that one of their dice is biased, showing the number 4 too frequently. The casino agree to test it by throwing the die 20 times. How many times would the number 4 need to come up to justify the customer's complaint at the 5% significance level?

13. A manufacturer of dog biscuits produces them in five different colours. Boxes are supposed on average to contain equal numbers of each colour. Samantha's dog prefers the black biscuits but Samantha suspects that there are fewer of them than there should be.

In one box of 50 biscuits there are only 3 black ones. Samantha complains that her dog is being deliberately deprived. Consider her complaint.

14. A railway line's management claims that 90% of the trains on its line arrive no more than 2 minutes after the scheduled arrival time. A regular user believes that the percentage achieving this standard is in fact lower.

Exercise 5A continued

She records her arrival time for her next 15 journeys and finds that the train is more than 2 minutes late on 3 occasions.

Does she have a justified complaint?

15. A psychologist claims to be able to tell if a person is telling a lie or the truth by watching someone's facial movements. His claim is tested by giving a volunteer twenty questions. The volunteer is asked to lie sometimes and to tell the truth sometimes, choosing the occasions at random. The psychologist, who can see the volunteer's face but not hear the answers, then has to judge whether the volunteer lied or told the truth.

The psychologist judged correctly 14 times out of 20.

Comment on the psychologist's claim.

16. Before I sat an examination, my teacher told me that I had a 60% chance of obtaining a grade A, but I thought I had a better chance than that.

In preparation for the examination, we did 7 tests each of the same standard as the examination. Assuming my teacher is right, find the probability that I would get grade A on
(i)   all 7 tests,
(ii) exactly 6 tests out of 7,
(iii) exactly 5 tests out of 7.
In fact I got a grade A on 6 tests out of 7. State suitable null and alternative hypotheses and carry out a statistical test to determine whether or not there is evidence that my teacher is underestimating my chances of a grade A.

17. A large number of people, 75% of them women, apply for 20 identical jobs. The employer finally selects 8 men and 12 women to fill the vacancies. There is much disappointment from those who are not chosen, and both the men and the women among them claim that they were the victims of gender discrimination. Is either group justified in its complaint?

1-tail and 2-tail Tests

Think back to the two examples in the first part of this chapter.

What would Dan have said if his eight children had all been girls?
What would Leonora have said if the number 1 had not come up at all?

In both our examples the claim was not only that something was unusual but that it was so in a particular direction. So we looked only at one side of the distributions when working out the probabilities, as you can see in figure 5.3 a and b. In both cases we applied 1-tail tests. (The word tail refers to the shaded part at the end of the distribution.)

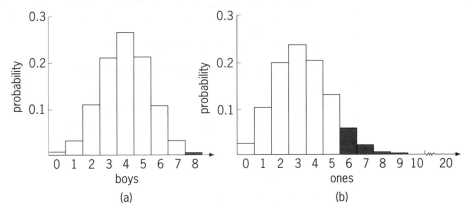

Figure 5.3

If Dan had just claimed that there was something odd about his chromosomes, then you would have had to work out the probability of a result as extreme on either side of the distribution, in this case 8 girls or 8 boys, and you would then apply a 2-tail test.

Here is an example of a 2-tail test.

EXAMPLE

The producer of a television programme claims that it is politically unbiased. "If you take someone off the street it is 50:50 whether he or she will say the programme favours the government or the opposition", she says.

However when ten people, selected at random, are asked the question "Does the programme support the government or the opposition?" nine say it supports the government.

Does this constitute evidence, at the 5% significance level, that the producer's claim is inaccurate?

Solution

Read the last sentence carefully and you will see that it does not say in which direction the bias must be. It does not ask if they are favouring the government or the opposition, only if the producer's claim is inaccurate. So you must consider both ends of the distribution, working out the probability of such an extreme result either way; 9 or 10 saying it favours the government, or 9 or 10 the opposition. This is a 2-tail test.

If p is the probability that someone believes the programme supports the government, you have

Null hypothesis, H_0	$p = 0.5$	Claim accurate
Alternative hypothesis, H_1	$p \neq 0.5$	Claim inaccurate
Significance level	5%	
	2-tail test	

The situation is modelled by the Binomial distribution $B(10, 0.5)$ generated by

$$(0.5 + 0.5t)^{10}$$

where the marker t indicates the number of people claiming bias towards the government. See figure 5.4.

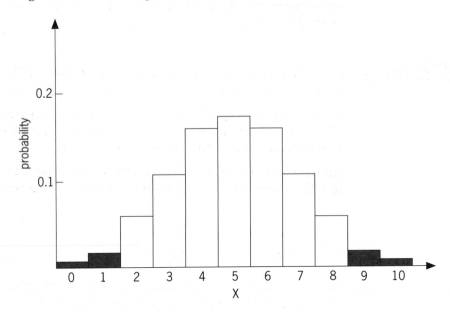

Figure 5.4

Expanding this expression gives

$$P(X = 0) = 1/1024 \qquad P(X = 1) = 10/1024$$
$$P(X = 10) = 1/1024 \qquad P(X = 9) = 10/1024$$

where X is the number of people claiming bias in favour of the government.

Thus the total probability for the two tails is 22/1024 or 2.15%.

Since 2.15% < 5% the null hypothesis is rejected in favour of the alternative, *that the producer's claim is inaccurate.*

N O T E

You have to look carefully at the way a test is worded to decide if it should be 1-tail or 2-tail.

Dan Ellis claimed his chromosomes made him more likely to father boys than girls. That requires a 1-tail test.

In Calvin's story, Leonora claimed the die was biased in the direction of too many 1s. Again a 1-tail test.

The test of the television producer's claim was for inaccuracy in either direction, and so a 2-tail test was needed.

Unsymmetrical Cases

In the example above the distribution was symmetrical and so the 2-tail test was quite simple to apply. In the next case the distribution is not symmetrical and the test has to be carried out by finding out the critical regions at each tail.

EXAMPLE

Pepper moths occur in two varieties, light and dark. The proportion of dark moths increases with certain types of atmospheric pollution.

In a particular village, 25% of the moths are dark, the rest light. A biologist wants to use them as a pollution indicator. She traps samples of 15 moths and counts how many of them are dark.

For what numbers of dark moths among the fifteen can she say, at the 10% Significance level, that the pollution level is changing?

Solution

In this question you are asked to find the Critical region for the test:

H_0 $p = 0.25$ (the proportion of dark moths is 25%)
H_1 $p \neq 0.25$ (the proportion is no longer 25%)
           Significance level 10%
           2-tail test

where p is the probability that a moth selected at random is dark.

You want to find each tail to be as nearly as possible 5% but both must be less than 5%, something that is easiest done using cumulative binomial distribution tables.

Looking under $n = 15$, for $p = 0.25$

n	x	0.050	0.100	0.150	1/6	0.200	0.250	0.300	1/3	0.350	0.400
15	0	0.4633	0.2059	0.0874	0.0649	0.0352	0.0134	0.0047	0.0023	0.0016	0.0005
	1	0.8290	0.5490	0.3186	0.2596	0.1671	0.0802	0.0353	0.0194	0.0142	0.0052
	2	0.9638	0.8159	0.6042	0.5322	0.3980	0.2361	0.1268	0.0794	0.0617	0.0271
	3	0.9945	0.9444	0.8227	0.7685	0.6482	0.4613	0.2969	0.2092	0.1727	0.0905
	4	0.9994	0.9873	0.9383	0.9102	0.8358	0.6865	0.5155	0.4041	0.3519	0.2173
	5	0.9999	0.9978	0.9832	0.9726	0.9389	0.8516	0.7216	0.6184	0.5643	0.4032
	6	1.0000	0.9997	0.9964	0.9934	0.9819	0.9434	0.8689	0.7970	0.7548	0.6098
	7		1.0000	0.9994	0.9987	0.9958	0.9827	0.9500	0.9118	0.8868	0.7869
	8			0.9999	0.9998	0.9992	0.9958	0.9848	0.9692	0.9578	0.9050
	9			1.0000	1.0000	0.9999	0.9992	0.9963	0.9915	0.9876	0.9662
	10					1.0000	0.9999	0.9993	0.9982	0.9972	0.9907
	11						1.0000	0.9999	0.9997	0.9995	0.9981
	12							1.0000	1.0000	0.9999	0.9997
	13									1.0000	1.0000
	14										
	15										

From this you can see that the left hand tail includes 0 but not 1 or more; the right hand tail is 8 and above but not 7.

So the critical regions are less than 1 and more than 7 dark moths in the twenty. For these values she would claim the pollution level is changing.

N O T E   *This is really quite a crude test. The left hand tail is 1.34%, the right hand (1 − 0.9827) or 1.73%. Neither is close to 5%. This situation would be improved if you were to increase the sample size; 15 is a small number of moths on which to base your findings. However for large samples you would expect to use either the normal or the Poisson approximation to the binomial distribution; you will meet these in the next book of this series.*

Exercise 5B

As for Exercise 5A, in all these questions you should apply the checklist on page 121 to the hypothesis test.

1. To test the claim that a coin is biased, it is tossed 20 times. It comes down heads 12 times. Test at the 10% significance level whether this claim is justified.

2. A biologist discovers a colony of a previously unknown type of bird nesting in a cave. Out of the 16 chicks which hatch out during his period of investigation, 13 are female. Test at the 5% significance level whether this supports the view that the sex ratio for the chicks differs from 1.

3. People entering an exhibition have to choose whether to turn left or right. Out of the first twelve people, 9 turn left and 3 right. Test at the 5% Significance level whether people are more likely to turn one way than another.

4. Weather records show that on average one day in four is wet in April at a certain seaside resort, but local people write to their newspaper complaining that the climate is changing.

   A reporter on the paper records the weather for the next 20 days and finds that 10 of them are wet.

   Do you think the complaint is justified?

5. In a fruit machine there are 5 drums which rotate independently to show one out of 6 types of fruit each (Lemon, Apple, Orange, Melon, Banana and Pear). You win a prize if all 5 stop showing the same fruit. A customer claims that the machine is fixed; the lemon in the first place is not showing the right number of times. The manager runs the machine 20 times and the lemon shows 6 times in the first place. Is the customer's complaint justified at the 10% significance level?

Exercise 5B continued

6. A boy is losing in a game of cards and claims that his opponent is cheating.

   Out of the last 12 times he shuffled and dealt the cards, the first card to be laid down was a spade on only one occasion. Can he justify his claim at the 2% significance level?

7. A small colony, 20 strong, of a previously unknown animal is discovered. It is believed it might be the same species as one described by early explorers who said that one quarter of them were male, the rest female.

   What numbers of males and females would lead them to believe, at the 5% significance level, that they are not the same species?

8. A multiple choice test has 20 questions, with the answer for each allowing four options, *A*, *B*, *C* and *D*. All the children in a class tell their teacher that they guessed all 20 answers. The teacher does not believe them. Devise a 2-tail test at the 10% significance level to apply to a child's mark to test the hypothesis that the answers were not selected at random.

9. When a certain language is written down, 15% of the letters are Z. Use this information to devise a test at the 10% significance level which someone who does not know the language could apply to a short passage, 50 letters long, to determine whether it is written in the same language.

10. A seed firm states on the packets of bean seeds that the germination rate is 80%. Each packet contains 25 seeds.
    (i)   How many seeds would you expect to germinate out of one packet?
    (ii) What is the probability of exactly 17 germinating?
    A man buys a packet and only 12 germinate.
    (iii) Is he justified in complaining?

KEY POINTS

Steps for conducting a hypothesis test:

- Establish the null and alternative hypotheses.
- Decide on the significance level.
- Collect suitable data. Items must be chosen independently and at random.
- Conduct the test.
- Interpret the result.
- Conduct the rest.
- Interpret the result.

Answers to selected exercises

1 EXPLORING DATA
Exercise 1C
1. (a) Mode = 45, mean = 39.6, midrange = 37, median = 41
   (b) Mode = 116 and 132 (bimodal), mean = 122.5, midrange = 124, median = 122
   (c) Mode = 6, mean = 5.3, midrange = 5, median = 6
2. (a) Mode = 14y 8m, mean = 14y 5.5m, midrange = median = 14y 6m
   (b) Mode = 4, mean = 3.85, midrange = 3.75, median = 4
   (c) Mode = 0,4 (bimodal), mean = 2.5, midrange = 4, median = 3
   (d) Mode = 0, mean = 52.83, midrange = 50, median = 58
   (e) Mode = 0 and 21 (bimodal), mean = 29.4, midrange = 84.5, median = 20
   (f) No clear mode, mean = 3.45, midrange = 3.5, median = 3.5
3. (i) Mode = 5, 6, mean = 5, midrange = 5, median = 5
   (ii) Mode = 50, 60, mean = 50, midrange = 50, median = 50
   (iii) Mode = 15, 16, mean = 15, midrange = 15, median = 15
   (iv) Mode = 10, 12, mean = 10, midrange = 10, median = 10
4. (i) Mode = 1, mean = 1.353, midrange = 5, median = 1
5. Mode = 1, mean = 2.089, midrange = 3.5, median = 2
6. 36.87 years, 35.59 years
7. (ii) 264 seconds
8. (a) 11.42 minutes (11 minutes 25 seconds)
   (b) 65 minutes
9. Mean mass = 138.9 g
10. Mean length = 42.6 cm

Exercise 1D
1. (a) (i) 7 (ii) 6 (iii) LQ = 4.5, UQ = 7.5 (iv) 1.5
   (b) (i) 15 (ii) 11 (iii) LQ = 8, UQ = 14 (iv) 3
   (c) (i) 10 (ii) 26 (iii) LQ = 24, UQ = 28 (iv) 2
   (d) (i) 29 (ii) 119.5 (iii) LQ = 115.5, UQ = 126 (iv) 5.25
2. (a) (i) 8 (ii) 5 (iii) LQ = 2.5, UQ = 7.5 (iv) 2.5
   (b) (i) 8 (ii) 15 (iii) LQ = 12.5, UQ = 17.5 (iv) 2.5
   (c) (i) 8 (ii) 275 (iii) LQ = 272.5, UQ = 277.5 (iv) 2.5
   (d) (i) 80 (ii) 50 (iii) LQ = 25, UQ = 75 (iv) 25
3. (i) 74 (ii) LQ = 73, UQ = 76 (iii) 1.5
5. (iii) Median = 65, IQR = 14 (estimated)
   (iv) Median = 66, IQR = 14
6. (ii) (a) Median = 11.25 secs (b) IQR = 11.8 secs (iii) mean = 13.1
7. (a) 17% (b) 54 g (c) 7.5 g
8. (b) 233 thousand (d) 2.2 tonnes (e) 516 thousand tonnes
9. Mean = 64.09 years, median = 66.5 years, IQR = 19.0 years
10. Probability = $0.46 \times 0.41 = 0.19$
11. (ii) Semi-IQR values: manual men = £53, manual women = £29.30, non-manual men = £91.50, non-manual women = £58.50
    (iii) Coefficient values: manual men = 0.23, manual women = 0.21, non-manual men = 0.28, non-manual women = 0.28

Exercise 1E
1. (a) (i) 5.5 (ii) 2.16
   (b) (i) 11.2 (ii) 3.93
   (c) (i) 25.8 (ii) 2.99
   (d) (i) 120.3 (ii) 7.89

2. (a) (i) 4.7 (ii) 2.7
   (b) (i) 14.7 (ii) 2.7
   (c) (i) 274.7 (ii) 2.7
   (d) (i) 46.7 (ii) 26.7
3. Mean = 7.7, standard deviation = 5.39
4. (i) Mean = 3, standard deviation = 1.506 (ii) 79%
5. Mean = 4.7, standard deviation = 2.52
6. (i) Mean = -4.3, standard deviation = 14
   (ii) Mean = 99.957 m, standard deviation = 0.14 m
   (iv) New mean = -2.05 cm, new standard deviation = 10.24 cm
7. (i) m = 12.2 secs, s = 6.11 secs (ii) 14.67
8. Mean = 6.5, standard deviation = 2.1 (30 games); mean = 6.8, standard deviation = 1.9 (50 games)
9. (i) 3147.72 g (ii) 84509.99
   (iii) n = 200, Σx = 5164.84, Σx^2 = 136549.91 (iv) 25.8242, 3.98
10. Monday's sum of squares = 6780, Tuesday's sum of squares = 4625. Combined standard deviation = 2.2
11. Mean = 6 volts, standard deviation = 0.067 volts
12. (ii) Neverlate Ltd: mean 10.1 days, SD = 0.700 days
    Ontime Ltd: mean = 10.5 days, SD = 2.46 days
13. (ii) 10th percentile = 17.5 (roughly)
    (iii) Mean = 29.84, SD = 9.89, median = 28.3
14. Mean = 113.8, median = 115.1, SD = 14.1, S = 0.092

2 PROBABILITY

Exercise 2A
1. 66/534, assuming each faulty torch has only one fault.
2. (i) $\frac{1}{6}$ (ii) $\frac{3}{6}$ (iii) $\frac{3}{6}$ (iv) $\frac{3}{6}$
3. (i) $\frac{12}{98}$ (ii) $\frac{53}{98}$ (iii) $\frac{45}{98}$ (iv) $\frac{42}{98}$ (v) $\frac{56}{98}$ (vi) $\frac{5}{98}$
4. (i) $\frac{5}{2000}$ (ii) $\frac{1995}{2000}$ (iii) Lose £100, if all tickets sold
   (iv) 25 p (v) 2500
5. (a) 0.35 (b) They might draw (c) 0.45 (d) 40%
6. (i) $\frac{10}{20}$ (ii) $\frac{4}{20}$ (iii) $\frac{10}{20}$ (iv) $\frac{2}{20}$ (v) $\frac{12}{20}$ (vi) 0 (vii) 1
   (viii) $P(E \cup S) = P(E) + P(S) - P(E \cap S)$
   (ix) $P(E \cup O) = P(E) + P(O) - P(E \cap O)$
7. (i) 0.22 (ii) 0.40 (iii) 0.10 (iv) 0.65 (v) 0.35

Exercise 2B
1. (i) 0.2401 (ii) 0.5002 (iii) 0.4998
2. (i) 0.0084 (ii) 0.1732 (iii) 0.1816
3. $\frac{1}{12}$
   (a) (i) 0.00058 (ii) 0.77 (iii) 0.020
   (b) (i) 0.0052 (ii) 0.52 (iii) 0.094
4. 0.93
5. (i) 0.0016 (ii) 0.0064 (iii) 0.2064 (iv) 0.7936
6. $\frac{3}{4}$
7. (i) 0.2436 (ii) 0.7564 (iii) 0.2308 (iv) 0.4308
8. (i) 0.0741 (ii) 0.5787 (iii) 0.5556
10. 0.5833
11. (ii) $\frac{3}{36}$ (iii) 7
12. (i) $\frac{1}{2}$ (ii) $\frac{1}{8}$ (iii) $\frac{3}{8}$
    (iv) lose $\frac{1}{4}$ of stake, i.e lose 37.5 p
13. 0.31
14. (i) 0.1667 (ii) 0.1389 (iii) 0.1157 (iv) 0.5787
15. (i) 0.0741 (ii) 0.9630 (iii) 0.0000508
16. (i) 1 (ii) 1 (iii) 1:1

Exercise 2C

1. (i) 0.6 (ii) 0.556 (iii) 0.625 (iv) 0.047 (v) 0.321 (vi) 0.075 (vii) 0.028 (viii) 0.0022 (ix) 0.00095 (x) 0.48
2. (a) (i) 0.160 (ii) 0.120 (iii) 0.400 (iv) 0.160 (v) 0.003 (vi) 0.088 (vii) 0.018
3. (i) $^{35}/_{100}$ (ii) $^{42}/_{100}$ (iii) $^{15}/_{65}$
4. (i) $^1/_6$ (ii) $^5/_{12}$ (iii) $^2/_5$
5. (i) $^7/_{15}$ (ii) $^{11}/_{21}$ (iii) $^5/_{21}$ (iv) $^5/_{11}$
6. (i) 30 (ii) $^7/_{40}$ (iii) $^7/_{10}$ (iv) $^7/_{15}$
7. (a) $^7/_{10}$ (b) $^{19}/_{40}$ (c) $^{10}/_{19}$
8. (i) 0.497 (ii) 0.477 (iii) 0.222 (iv) 0.461
9. (i) 0.5 and 0.875
10. (i)

	Hunter dies	Hunter lives	
Quark dies	$^1/_{12}$	$^5/_{12}$	$^1/_2$
Quark lives	$^2/_{12}$	$^1/_3$	$^1/_2$
	$^1/_4$	$^3/_4$	1

(ii) $^1/_{12}$ (iii) $^5/_{12}$ (iv) $^5/_6$

3 SELECTIONS

Exercise 3A

1. (i) 40 320 (ii) 56 (iii) $^5/_7$
2. (i) $^1/_n$ (ii) $n-1$
3. (i) $(n+3)(n+2)$ (ii) $n(n-1)$
4. (i) $\dfrac{8! \times 2!}{5! \times 5!}$ (ii) $\dfrac{16!}{14! \times 4!}$ (iii) $\dfrac{(n+1)!}{(n-2)! \times 4!}$
5. (i) $9 \times 7!$ (ii) $n!(n+2)$
6. 24
7. 360
8. 40 320
9. 720
10. $n + m$
11. $n \geq 7$, $n \geq 8$, $n \geq 9$ where $n \in \mathbb{N}$

Exercise 3B

1. 182
2. 280
3. 1665
4. (i) 60 (ii) 125
5. 729
6. 24, 3024, 6561, 9000
7. (i) 0.096 (ii) 0.482
8. 120
9. 25

Exercise 3C

1. (i) 360 (ii) 90 720
2. 50 400
3. (i) 45 360 (ii) 1260
4. 1512
5. (i) 0.417 (ii) 0.083
6. $^1/_9$
7. 120
8. (i) 13 (ii) 26
9. 34 560
10. 36

Exercise 3D

1. 120
2. (i) 72 (ii) 1875
3. $^6/_{11}$
4. 13 776
5. 270 725, 20 825
6. $^7/_{32}$
7. 38 760, 0.0238
8. (i) 360 (ii) 72
9. 3 628 800, 28 800, 28 800
10. (i) 0.0198 (ii) 0.385
11. 6188, 1287, 462, 11!, 2 × 10!

Exercise 3E

1. (i) 3 628 800 (ii) 25 200
2. (i) $(k + 5)(k + 4)$ (ii) $(r + s + 1)$
3. 24
4. 720
5. (i) 5040 (ii) 226 800
6. 978
7. 0.0189
8. 72
9. (i) 18 (ii) 192
10. 179 335
11. (i) 715 (ii) 245
12. 2187
13. 0.311
14. (i) 31 824 (ii) 3000 (iii) 600 (iv) 4896
15. 120 (i) 4 (ii) 60 (iii) 20, 4 left, 0.5
16. 210; (i) $^1/_{14}$ (ii) $^3/_7$ (iii) $^1/_7$ (iv) $^4/_{15}$
17. (i) 16 (ii) 14
18. (i) 105 (ii) 11 025 (iii) 8 (iv) − (v) 1080

4 THE BINOMIAL DISTRIBUTION
Exercise 4A

1–10 Coefficients

1. 1   2   1
2. 1   3   3   1
3. 1   4   6   4   1
4. 1   5   10   10   5   1
5. 1   6   15   20   15   6   1
6. 1   8   28   56   70   56   28   8   1
7. $^1/_{16}$ $^1/_4$ $^3/_8$ $^1/_4$ $^1/_{16}$
8. 0.402   0.402   0.161   0.032   0.003   0.0001
9. 0.3966   0.3966   0.1652   0.0367   0.0046   0.0003   0.000008
10. 0.1335   0.3115   0.3115   0.1730   0.0577   0.0115   0.0013   0.00006
11. $^3/_8 h^2 t^2$

12. $\dfrac{40}{243} n^2 y^3$

13. $0.2009 n^4 s^2$
14. $0.2086 t^3$
15. $0.0001714 t^5$

Exercise 4B

1. $^{15}/_{64}$
2. 0.271

3. 0.214
4. 0.294
5. 0.146
6. (i) $1/8$ (ii) $3/8$ (iii) $3/8$ (iv) $1/8$
7. (i) 0.246 (ii) Exactly 7
8. (i) 0.058 (ii) 0.198 (iii) 0.296 (iv) 0.448 (v) 2
9. (i) 0.264 (ii) 0.368 (iii) 0.239 (iv) 0.129
11. (i) 0.000129 (ii) 0.0322 (iii) 0.402 (iv) 0 & 1 equally likely
12. (i) 0.268 (ii) Less than 2
13. (i) 2 (ii) 0.388 (iii) 0.323
14. (i) 0.240 (ii) 0.412 (iii) 0.265 (iv) 0.512 (v) 0.384 (vi) 0.096 (vii) 0.317
15. (i) Yes (ii) 2.5 (iii) 2
16. (i) $1/81$ (ii) $8/81$ (iii) $24/81$ (iv) $32/81$ (v) 2 min 40 secs
17. (ii) 0.00293 (iii) At the bar (iv) 0.242 (v) 2.71
18. (i) 0.0735 (ii) 2.7 (iii) 0.267
19. (i) $32/243$ (ii) $192/243$ (iii) $40/243$ (iv) $242/243$
20. (i) 0.0322 (ii) 0.0355 (iii) 0.00129 (iv) 0.04
21. (i) 0.00269 (ii) 0.121 (iii) 0.167 (iv) 0.723
22. (i) 0.195 (ii) 0.299 (iii) 0.260
23. (i) $10/32$ (ii) $13/16, 8/27$
24. (i) 0.003 22 (ii) 0.0193 (iii) 0.0386
25. (i) 0.349 (ii) 0.387 (iii) 0.194 (iv) 0.070 (v) 0.678

5 HYPOTHESIS TESTING USING THE BINOMIAL DISTRIBUTION

Exercise 5A

1. 0.1275   Accept H_0
2. 0.016   Reject H_0
3. (i) 0.180 (ii) 0.416 $0 \leqslant N \leqslant 7$
4. 0.0106   Reject H_0
5. For 8 years ('83 – '91); H_0 p = ½, H_1 p > ½: 0.14   Accept H_0
6. 0.055   Reject H_0
7. 0.017   Reject H_1
8. 0.240   Accept H_0
9. 0.048   Reject H_0
10. 0.104   Accept H_0
11. 0.0867   Accept H_0 at 5% level
12. 7
13. 0.00566   Reject H_0 at 5% level
14. 0.184   Accept H_0 at 5% level
15. 0.057   Accept H_0 at 5% level
16. (i) 0.0280 (ii) 0.131 (iii) 0.261   Accept H_0 at 5% level
17. Complaints not justified at 10% level

Exercise 5B

1. 2 × 0.252   Accept H_0
2. 2 × 0.0106   Reject H_0
3. 2 × 0.073   Accept H_0
4. 0.014   Reject H_0 but data not independent
5. 0.102   Accept H_0
6. 0.158   Accept H_0
7. $\leqslant 1$ or > 9 males
8. $\leqslant 1$ or > 8 correct
9. Critical region is $\leqslant 3$ or $\geqslant 13$
10. (i) 20 (ii) 0.0623 (iii) Complaint justified